Contents

LIVING THINGS IN ACTION

Revised Nuffield
BIOLOGY
TEXT 2

Published for the Nuffield Foundation
by Longman Group Limited

Longman Group Limited
London
*Associated companies, branches, and representatives
throughout the world*

First published 1966
Revised edition 1975, reprinted 1977, 1978
Copyright © The Nuffield Foundation, 1966, 1975
ISBN 0 582 04602 5

Design and art direction by Ivan and Robin Dodd

Filmset in 11 on 12 point Century Schoolbook
by Photoprint Plates Ltd., Rayleigh, Essex
and made and printed in Great Britain
by Butler and Tanner Ltd., Frome and London

Foreword

It is ten years since the Nuffield Foundation undertook to sponsor curriculum development in science. The subsequent projects can now be seen in retrospect as forerunners in a decade unparalleled for interest in teaching and learning not only in but far beyond the sciences. Their success is not to be measured simply by sales but by their undoubted influence and stimulus to discussion among teachers—both convinced and not-so-convinced. The examinations accompanying the schemes of study which have been developed with the ready cooperation of School Certificate Examination Boards have provoked change and have enabled teachers to realize more fully their objectives in the classroom and laboratory. But curriculum development must itself be continuously renewed if it is to encourage innovation and not be guilty of the very sins it sets out to avoid. The opportunities for local curriculum study have seldom been greater and the creation of Schools Council and Teachers' Centres have done much to contribute to discussion and participation of teachers in this work. It is these discussions which have enabled the Nuffield Foundation to take note of changing views, correct or change emphasis in the curriculum in science, and pay attention to current attitudes to school organization. As always, we have leaned on many, particularly those in the Association for Science Education who, through their writings, conversations, and contributions in other varied ways, have brought to our attention the needs of the practising teacher and the pupil in schools.

This new edition of the Nuffield Biology *Texts* and *Teachers' guides* draws heavily on the work of the editors and authors of the first edition, to whom an immense debt is owed. The first edition, published in 1966, was edited by Professor W. H. Dowdeswell, organizer of the Biology project which carried out the trials in schools of the original draft materials. The authors of the first edition

were:
Alison Leadley Brown Texts I and II and Teachers' Guides I and II
C. D. Bingham Texts I and II and Teachers' Guides I and II
A.K. Thomas Text III and Teachers' Guide III
A. Ellis Texts III and IV and Teachers' Guides III and IV
A. Darlington Texts III and IV and Teachers' Guides III and IV
P. J. Kelly Text V and Teachers' Guide V

The new edition contains a preponderant part of these authors' material, either in its original form or in edited versions. They are credited among the authors of the new edition but their wider contribution in providing a firm basis for further developments must be gratefully acknowledged here.

I particularly wish to record our gratitude to Grace Monger the General Editor of this new series. It has been her responsibility to organize and coordinate this revision and it is largely through her efforts that we have been able to ensure the fullest cooperation between teachers and the authors.

As always I should like to acknowledge the work of William Anderson, our publications manager, and his colleagues, and, of course, to thank our publisher, the Longman Group, for continued assistance in the preparation and publication of these books. I must also record our debt to those members of Penguin Education who were actively involved in the preparation of the books until a late stage in their production. The contribution of editors and publishers to the work of the course team is not only most valued but central to effective curriculum development.

K. W. Keohane
Co-ordinator of the Nuffield Foundation Science Teaching Project

Preface
to the first edition

You have probably been told that biology is the study of living things. This is true but, by itself, the statement does not tell you what biology is really about. In this course we want to do more than just teach you how living things function; we want you to understand why scientists wish to know about life and how they set about finding out biological truths.

Scientists must be curious; they must be prepared to form tentative guessing answers to the questions they ask themselves; and they must be able to test these guesses. We have tried to guide you through this process, to show you why you should be curious, what kinds of questions you should investigate, and how you should devise and carry out experiments. Experiments are not intended to prove things you already know; they are to investigate whether something does or does not happen so that you can form hypotheses which, themselves, can be tested by further experiments. Thus, a negative result may be as important as a positive one. We have also tried to indicate how you should use the results you get; how you should test them further and how you should relate them to the questions you posed yourselves.

By the end of this course we hope you will know not only more about living things, particularly man, but also more about how to study living things both in the laboratory and in their natural state.

The essence of this Nuffield course in biology is 'science for all'. In devising it, our intention has been to provide a balanced and up-to-date view of the subject suitable for pupils who will leave school at the age of sixteen and do no more formal biology. For some it will also provide a jumping-off point for further study at A-level.

The course has been built around a number of fundamental themes. Such issues as the relationship of structure and function, adaptation, and the interaction of organism and environment are discussed again and again in different contexts throughout the five-year period. The course is designed to foster a critical approach to the subject with an emphasis on experimentation and enquiry rather than on the mere assimilation of facts. In terms of a conventional syllabus this means that less factual matter is included. This in itself is no bad thing, provided the principles of the new teaching are accepted and the methods used are in sympathy with the aims of the course. In order to foster this outlook, a *Teachers' guide* has been produced. This is closely cross-referenced to the *Text* and contains copious notes on teaching methods, also much technical matter relating to practical work, including additional experiments. In short, our aim has been to produce not only a new syllabus, but, more important, a new approach to teaching.

The course falls clearly into two parts; the first two years which can be regarded as introductory, and the remaining three which constitute the next (intermediate) phase. The introductory phase is characterized by a broad general approach to the subject. In the intermediate phase the treatment becomes more quantitative with greater emphasis on experimentation and reasoning.

Preface
to the second edition

The most important feature of the second edition of the Nuffield Biology *Texts* is the part played in their revision by teachers and pupils who have had the experience of following the course to O-level. Before any decisions were made to change the materials first published in 1966, a long process of evaluating how far they had succeeded in fulfilling the aims and intentions of the original Project took place. From this exhaustive investigation into the use of the first edition, the editors and authors of this new edition have drawn invaluable help in deciding how to present the material and to take into account the criticisms and suggestions of practising teachers. In this sense, the second edition can be seen as the result of a further stage of the trials on which the original materials were based.

The introductory phase (two years) is now in one volume instead of two but the intermediate phase is still covered in three volumes. The subject matter has undergone considerable rearrangement and re-editing. Where topics have been developed at greater length in some cases, or have been shortened in others, these changes are based on the reports and requests from schools which were received during the evaluation described above.

The intentions of the course, however, remain the same and the aim of the revision has been to refine the materials and to bring them up to date where this is necessary.

Grace Monger, General Editor

Size and support

When you see something which you find interesting, you may ask yourself how it happened or how it works. You might suggest an answer and then go on to test your suggestion to see whether it is a good one. This is what a scientist does when he tests his ideas by means of an experiment. To produce reliable results, his experiments must be set up very carefully, so that he knows exactly what is being tested. A good experiment is one in which as many as possible of the conditions affecting it are known and where only one of these is altered at a time.

Biological experiments are difficult to design because living things are so complicated and variable. Often, two sets of apparatus are used in which all the conditions *except one* are kept exactly the same. The results from the two sets can then be compared and the effect of altering this one condition can be seen more clearly. The second set of apparatus, in which all the conditions are as near normal as possible, is thus used as a standard and is called a *control*. If the same answer is obtained after the experiment has been repeated several times, it is more likely to be reliable. If, however, your results do not support your original idea, then you may have to change it and start again.

When looking at 'living things in action' (the title of this book) you will be seeking answers to a number of problems. You may find some possible solutions from experiments but you will need to search for evidence from a variety of sources. Your answers may not always be clear but, if you check them carefully against each other, you will gradually build up a much fuller understanding of how organisms work.

1.2 The problem of size and support in mammals

You can go a long way toward solving this problem by looking at mammals of various shapes and sizes (*figures 1* and *2*) and asking yourself how they hold themselves up.

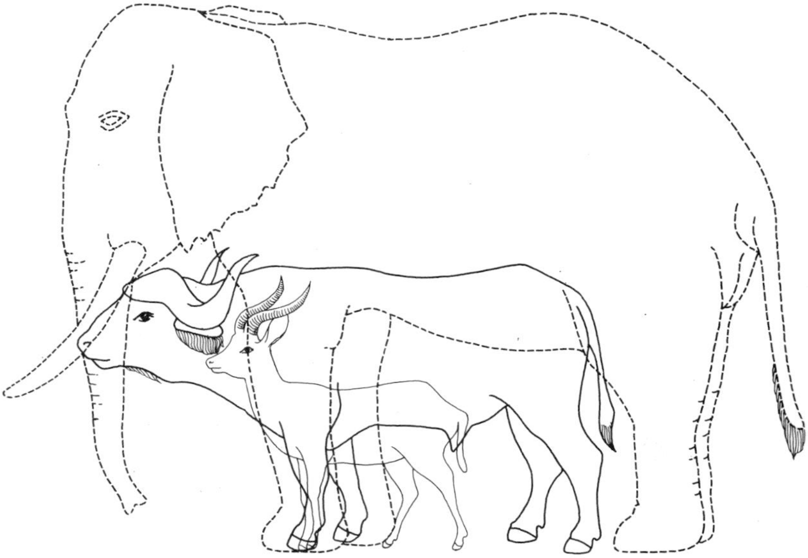

Figure 1
How do these three mammals of varying shapes and sizes hold themselves up?

	Mass (kg)	Shoulder height (cm)
African elephant (*Loxodonta africana*)	6600	320
Cape buffalo (*Syncerus caffer caffer*)	820	170
kob (*Adenota kob*)	90	100

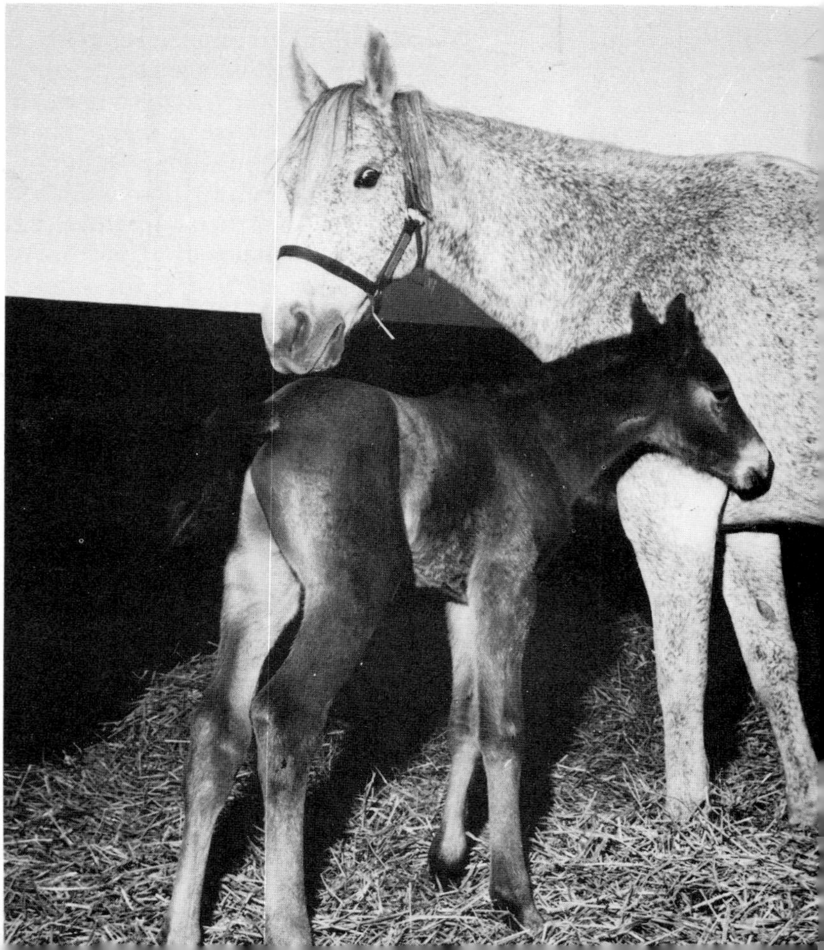

Figure 2
A mare and foal. What are the problems of support for this one species at different stages?
Photograph, Racing Information Bureau.

2

Figure 3
These pictures of a mole (**a**),
a weasel (**b**), and seals (**c**)
illustrate how the shape of a living
thing is related to its way of life.
Note that in spite of their wide
differences, **a** and **b** are much the
same in mass.

When considering size and support, remember that the
shape of a living thing is related to its way of life;
similarly the shape of an organ of the body is related to
the functions it has to perform. Legs, for instance, have a
number of functions. One function may be support, in
which the legs act as struts to keep the body up, while
another function may be movement, in which the legs act
as levers to propel the body along.

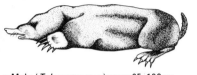

a Mole (*Talpa europaea*) mass 65–120 g;
   body length 11.5– 15 cm

b Weasel (*Mustela vulgaris*) mass 60–130 g;
   body length 21–23 cm

c Seals (*Halichoerus grypus*)
   mass of male animal about 280 kg;
   body length 3 m

Q1   Look at *figure 3*. Where do each of the animals live most
     of the time?

Q2   What do you think the limbs of each animal are used for?

You can see that we are up against the difficulty of trying
to study one function on its own when, in a living animal,
a limb may perform several functions, sometimes at the
same time.

Look carefully at *figures 1, 2, and 3* and as many other
pictures of mammals as you can get.

Q3   What do you notice about the length and thickness of
     their legs in relation to the size of the animal?

In descriptions of mammals, measurements of height to
shoulder, length of body from nose tip to base of tail, and
length of tail are usually given as well as the mass of the
body.

Length, thickness, and height are all linear measurements and nowadays will be given in centimetres, although some books will record them in inches (1 inch = 2.54 cm). Mass will usually be given in kilograms but again, some books will use pounds (1 kg = 2.205 lb). The mass of an animal refers to the amount of bone, muscle, and other tissues; in fact, to all the stuff of which an animal is made. Volume (in cm^3) could also be used to give the idea of the amount of stuff making up the size of an animal.

Because you are studying the problem of support, you will want information on body mass and, in the case of land-living mammals, you will need information on length of legs and their thickness or diameter. Some of this information is given in *table 1*.

Name	Mass (kg)	Shoulder height (cm)
camel	450	230
Cape buffalo	820	170
cat	2.5	18
cheetah	21	77
eland (antelope)	900	200
elephant	6600	320
giraffe	1200	490
greyhound	13	73
horse	650	175
hippopotamus	1700	145
lion	100	100
mouse	0.028	3
shrew	0.017	1.5
white rhinoceros	4064	175

Table 1

An elephant is a very big animal (6600 kg, shoulder height 320 cm) and it has big thick legs.

A mouse is a very small animal (0.028 kg, shoulder height 3 cm) and it has small thin legs.

Imagine a mouse enlarged until its body is as big as an elephant's.

Q4 Would its legs support its body if they remained in the same proportion as in the original mouse?

Q5 What evidence can you provide in support of your answer?

This is where models are helpful.

Look at the cube models in *figure 4*. Assume that models *a*, *b*, and *c*, are all built on the same plan. This means that the legs must always remain in the same proportion to the body. Using small wooden cubes you could try to work out the answers to the questions that follow.

Figure 4
A series of cube models to show
the effect of doubling the linear
measurements.

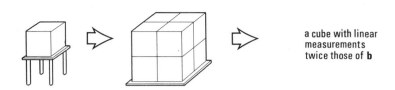

a cube with linear
measurements
twice those of **b**

Q6 *1* What has been done to the *linear* measurements of cube *a*
to make it into cube *b*?
2 How many times longer will the legs of *b* be than those
of *a*?
3 By how much will the thickness of each leg of *b* be
increased compared with *a* and how many straws will be
needed for each one?
4 How many small cubes are needed to build *b* and how
many times larger is the volume of *b* than the volume of *a*?

Q7 If the *linear* measurement of *c* is twice that of *b*:
1 How many times longer will the legs of *c* be than those
of *b*?
2 By how much will the thickness of each leg be increased
compared with *b* and how many straws will be needed for
each leg of *c*?
3 How many small cubes will be needed to make *c* and how
many times larger is the volume of *c* than the volume of *b*?

Each small cube will have a mass which you could find by
weighing it. Assuming that each has the same mass, you
will see from the answers to questions 6 and 7 that you
could find the mass of each of cubes *a* and *b* and *c*. This is
easily done by multiplying the mass of one small cube by
the number of small cubes making up the whole. Think
about this and then answer question 8.

Q8 When a bigger model is made on the same plan as a smaller
one, which of the following statements do you agree with?
a The mass of the body and the thickness of the legs
increase at the same rate.
b The mass of the body increases more rapidly than the
thickness of the leg increases.
c The mass of the body increases less rapidly than the
thickness of the leg increases.

Support whichever statement you choose by using the actual figures you have worked out in answer to questions 6 and 7.

You may now be able to decide whether big animals could be built to the same proportions as small animals.

You could test your idea by using a set of containers which will be provided. For an accurate answer you should know the relationship of each container to the others. You can find out if they are in proportion by measuring one characteristic of each model and comparing them. Length of leg would be an easy one to measure. It would also be useful to know the volume of each container as you are considering support and size. You can discover this by removing the containers from their stands and seeing how much sand or water would be needed to fill each of them. The containers on their stands supported by straw legs represent land-living mammals which normally have four legs but often use only three at any one time. Your models can therefore have either three or four legs. With the containers back in position on their legs, you can now find out whether in each case the legs will support a full container of sand or water. To compare results you may wish to measure the amount of sand or water at the point of collapse, so stand the models on a plastic tray in case you spill any from inside the containers.

Q9 Is there a limit to the size your 'container' animals could reach with their legs and the rest of their bodies remaining in proportion?

Q10 Using your results and any evidence from pictures, living mammals, and the data in *table 1*, make suggestions for suitable alterations to the legs of the stands to increase their efficiency.

If you can, test your modifications experimentally.

If you think about the work you have just done with the container stands, you will realize that it is the mass of the animal's body which is the important factor in deciding whether the legs can support it or not. Volume may be important too, for example, because a very large animal may find it difficult to move quickly when in danger. But it is because larger animals are usually also heavier that size and support are related. If, however, one animal has a volume ten times greater than another, you can only assume that its mass will be ten times greater if both are made of the same material.

1.3 Other problems of support in mammals

Sometimes it is interesting to work out things for yourself without being told exactly what to do. Here are some questions for you to think about and try to come to some conclusions about.

'Pillars' or 'zig-zags'
Look at the arrangement of the bones of the fore- and hind limbs of the horse and elephant (*figure 5*). Compare the masses of the elephant and the horse and decide whether each is linked with the arrangements of the bones.

Figure 5
The arrangement of bones in the limbs (**a**) of the horse, (**b**) of the elephant.

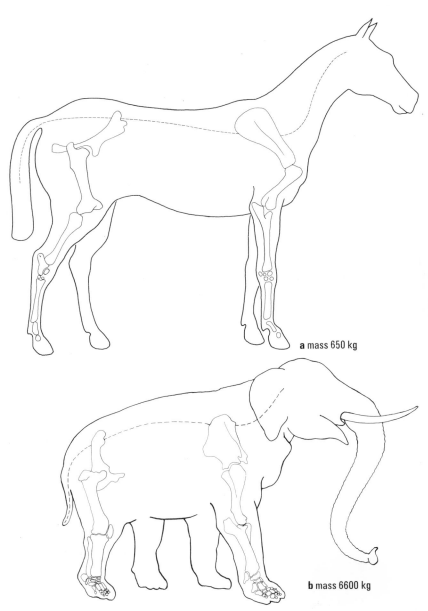

a mass 650 kg

b mass 6600 kg

Q1 Explain whether you think that 'pillars' or 'zig-zags' are better for supporting an enormous mass.

Q2 From the data in *table 1*, which mammals would you expect to have the same arrangement as the elephant and which will be more like the horse?

Later on, you will see how this arrangement fits in with movement because legs are not used for support only.

Could a whale have legs?
Going back to the experiment you did using the three differently sized containers, look again at your results and the suggestions you made to modify the legs.

Q3 Could a whale (mass 100 000 kg) have legs which would both support a body as heavy as this on land while allowing the whale to walk? It would be interesting to plan the size which the legs might have to be and then asking yourself if they would be much good to the whale!

Supporting soft internal organs
If you lie flat on the ground face downwards, you can find out what your chest and abdomen feel like if you keep your arms well out of the way and don't let them support you. Consider the support of soft organs in animals as big as whales. You may be able to experiment with a rubber balloon which has been filled with water instead of air.

1 Look at the shape of the balloon when it is placed in an aquarium full of water.
2 Put both hands carefully under the balloon, lift it out of the water, and put it on the bench or on clear Perspex.
3 Look at the shape of the balloon as it rests on the bench or on Perspex, and compare its shape now, with what you remember of its shape when it was in water.
4 Support the balloon of water in a sling and hook this onto a weighing scale.

Record the mass of the balloon as it hangs from the scale.

5 With the balloon still hanging in its sling from the scale, lower it into the water and watch the pointer on the scale.

Q4 Using the results from the experiment with the balloon, explain how the whale's way of life helps to prevent its soft internal organs from being crushed.

When you get down on all fours can you feel your abdomen sag a little? Do you feel your chest sag as much?

A moment's thought and a look at the diagram of the human skeleton (*figure 18,* page 19) will give a reason.

Figure 6 illustrates mammals representing groups of three different sizes and gives their masses.

Q5 How does the construction of their skeletons and that of the human skeleton (*figure 18,* page 19) fit in with what you have just been reading and doing?

Figure 6
Skeletons of animals in three different groups of sizes.

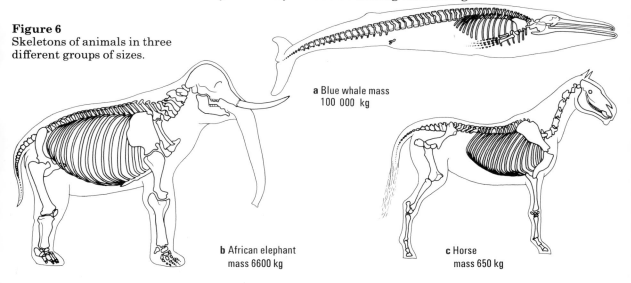

a Blue whale mass
100 000 kg

b African elephant
mass 6600 kg

c Horse
mass 650 kg

1.4 The problem of size and support in plants

Like animals, plants vary considerably in size and shape. In order to carry out many of their essential processes they must remain erect, so they must be able to support themselves in order to survive. You can try to find out how they do it. The great majority of plants are rooted in soil, so you might start by looking carefully at the roots and shoots of any land plants you can find and are allowed to dig up. Keep your own detailed record of what you see and do, if it is relevant to the problem you are studying.

Here are some ideas, but if you have others, use them too.
1 Try pulling up a variety of common weed plants such as daisy, plantain, dock, dandelion, buttercup, groundsel, shepherd's purse, and a small grass plant.
Try to find at least one plant from each group.
2 Dig up some of these plants carefully so that their main roots are not spoilt. Find other plants which have a stem about 20 to 30 cm tall and dig these up as well.
3 Measure the spread of the root and its depth and compare these measurements with similar ones for the shoot.
4 As soon as you have dug up the plants, feel both the root and the shoot. How rigid and how flexible are they?

5   With a sharp knife, cut across a stem and a thick root and look carefully at the cut surfaces with a hand lens.
6   Cut some very thin slices from a shoot and a thick root and put them in a dye (this will stain the woody parts).
7   Wash the slices in water and choose your best slices of root and shoot to mount side by side on a microscope slide. Put a drop of water on the slices and look at them carefully with a hand lens or under the low power lens of a microscope. Look especially at the shape of the root and stem slices and the arrangement of the woody parts.
8   Now look at some bigger plants, such as trees (*figure 7*).

Q1   How are such large plants supported?

Wellingtonia
(*Sequoia gigantea*)
97 m high

Oak
(*Quercus robur*)
30 m high

Oil palm
(*Elaeis guinieënsis*)
14 m high

Figure 7
Drawings by Mary Thomas.

1.5 Collecting further evidence

Forces acting on plants

Q1   Look at *figure 8* and say what is responsible for the two forces in the drawing.

Q2   How do they affect the plant?

Try to check your answer to question 2, using a potato chip prepared as shown in *figure 9*.

1   Bend the chip sideways as far as you can without breaking it. Check the marks while you do so.
2   Try pressing down on the chip and record what happens.

Figure 8
Forces acting on plants.

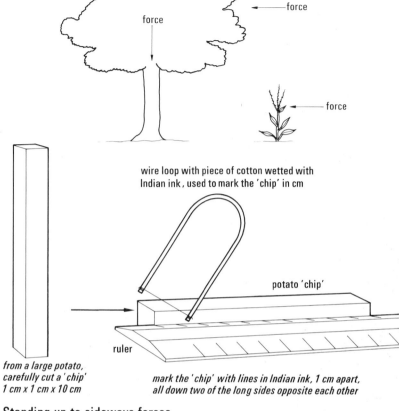

force

force

force

Figure 9
Preparing an experiment to
investigate the forces acting on
plants.

wire loop with piece of cotton wetted with
Indian ink, used to mark the 'chip' in cm

potato 'chip'

ruler

*from a large potato,
carefully cut a 'chip'
1 cm x 1 cm x 10 cm*

*mark the 'chip' with lines in Indian ink, 1 cm apart,
all down two of the long sides opposite each other*

Standing up to sideways forces

Test the resistance of a stem as described in *figure 10*.

Figure 10
An experiment to test how well
stems resist sideways forces.
For comparison, repeat it with a
drinking straw.

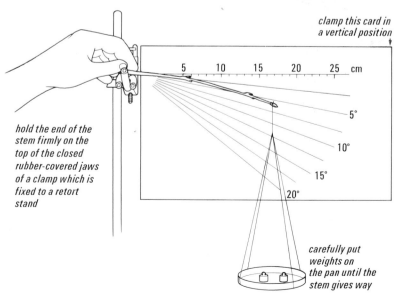

*clamp this card in
a vertical position*

5   10   15   20   25   cm

5°

10°

15°

20°

*hold the end of the
stem firmly on the
top of the closed
rubber-covered jaws
of a clamp which is
fixed to a retort
stand*

*carefully put
weights on
the pan until the
stem gives way*

Think about how you are going to make a fair comparison of strengths. Remember you must design your experiment so that you are measuring one factor only, that is, the force the stem can withstand before it gives way. When you have obtained your results look at the shape of the cut ends of the stems. Try to find the position of the strong supporting tissues. *Figure 11* may help you. It shows the detail of three different types of stems. As far as you can, explain how the arrangement of tissues and the shape of your stems affect their efficiency as supporting organs.

Figure 11
These drawings and the photographs in transverse section illustrate the structure of three types of plant stem.
Photographs, Teaching Media Centre, University of Southampton.

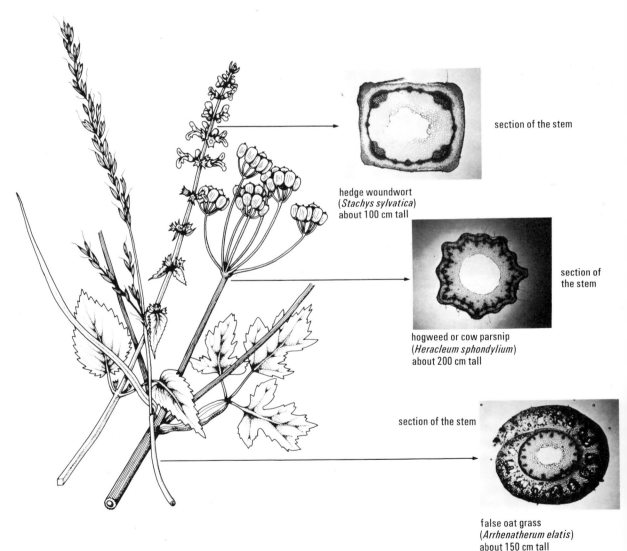

section of the stem

hedge woundwort
(*Stachys sylvatica*)
about 100 cm tall

section of the stem

hogweed or cow parsnip
(*Heracleum sphondylium*)
about 200 cm tall

section of the stem

false oat grass
(*Arrhenatherum elatis*)
about 150 cm tall

Living things in action

The effect of wind on different plants

A variety of experiments can be done with the arrangement shown in *figure 12*. Plants can be compared showing contrasting features such as:
thick and thin stems
bushy and slender growth form
leafy and non-leafy shoots
shoots with broad leaves and long thin leaves
a variety of differently constructed stems.

The effect of winds of different velocities can also be found using the same or similar plant material.

Figure 12
Apparatus for testing how far different types of plant are bent by the wind.

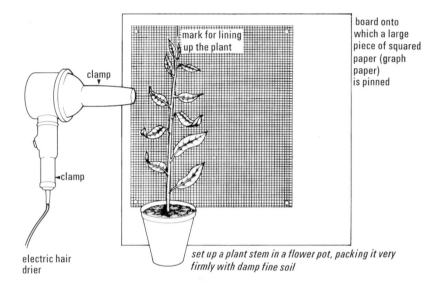

mark for lining up the plant

clamp

board onto which a large piece of squared paper (graph paper) is pinned

clamp

electric hair drier

set up a plant stem in a flower pot, packing it very firmly with damp fine soil

Stability at the base

See *figure 13* on the next page.

The paper cone and the paper cylinder are standing side by side and an electric hair drier is held some distance from them. It is gradually brought nearer until one, or the other, or both blow over. Repeat the investigation at least three times to check your observations – measure the distance from the hair drier to the paper shapes each time. Consider your results and see if you can apply them to the shapes of tree trunks.

Figure 13
Which of these paper models
stands up to the wind better?

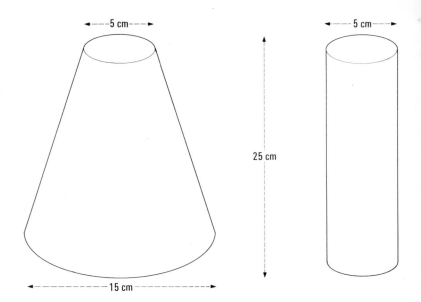

1.6 Animals, plants, and man-made structures

Plant structure is determined by fairly simple physical
laws which apply equally to animals and to man-made
structures. Look at *figure 14a* (the Eiffel Tower) and *b* (the
foot and ankle), and also at the Wellingtonia tree in
figure 7 (page 10).

Figure 14
Compare the physical laws
operating in the structure of the
tower and the foot and in that of
the Wellingtonia tree in *figure 7*.

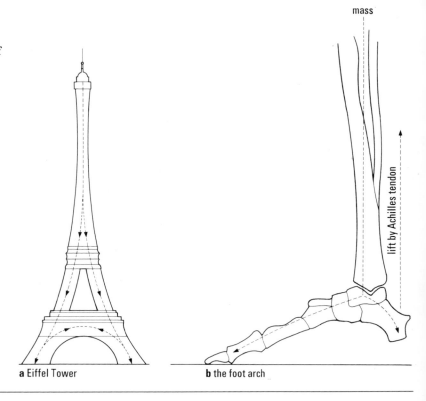

a Eiffel Tower   **b** the foot arch

Living things in action

Q1   What similar features do these structures have?

Q2   In what ways do they differ?

Q3   Can you suggest any reasons for the similarities and differences which you find?

The apparatus in *figure 15* might help you to see if your ideas are sensible.

Figure 15
One kind of apparatus for testing shape and stability.

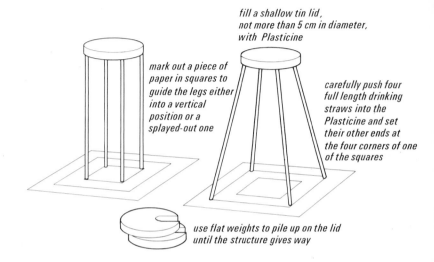

fill a shallow tin lid, not more than 5 cm in diameter, with Plasticine

mark out a piece of paper in squares to guide the legs either into a vertical position or a splayed-out one

carefully push four full length drinking straws into the Plasticine and set their other ends at the four corners of one of the squares

use flat weights to pile up on the lid until the structure gives way

1.7 The part played by water in the support of plants

There are two ways in which water may help a plant to remain upright. One you can see if you look first at a water weed growing in an aquarium and then when it is removed and held at the base only, as if it were rooted. You can investigate the second way in which water helps, by the following method.

1   Take two similar, flexible shoots of a non-woody land plant.
2   Leave one – shoot *a* – with its cut end in a jar of water.
3   Leave the second – shoot *b* – out on the bench without water for an hour.
4   At the end of the hour look at both stems and feel them.

Q1   Where is the water normally in relation to the water weed and the land plant?

Q2   Will the water content of shoot *b* be the same at the end of the hour as at the start of it?
Give reasons for your answer.

Q3   Account for the appearance and feel of shoots *a* and *b* at the end of the experiment by describing what has happened inside their stems.

1.8 Applying ideas

Here are two items concerning support in plants which will help you to see if you can apply the ideas which you have gained from the work you have been doing.

Getting better yields from wheat
Wheat has heavy ears of grain and only too frequently rain and wind can ruin the nearly ripe crop by breaking the stem, flattening the plants, and making it impossible for combine harvesters to operate. The grain is the most important part, and wheats have been bred to give greater numbers of grain per ear. As plants producing heavier heads have been developed, the plant breeders have had to consider the development of a stem which could support the extra mass.

Q1 Suggest what stem characteristics the plant breeder would be looking for.

Plant growth: form and habitat
Look at the drawings of the two plants shown in *figure 16*. One has been growing in a hedgerow and one on open wasteland.

Q2 Which plant do you think grew in which habitat? Explain your evidence.

Figure 16
Two specimens of goosegrass (*Galium aparine*), one of which was growing on open wasteland and one in a hedge.

2

Animal movement

2.1 Animals in motion
In Chapter 1 limbs were considered chiefly as supports, but they have other functions as well. One of the most important of these functions is to help an animal to move. Animals vary enormously, both in the way they move and the speed at which they move. You will realize this if you compare an eagle with an eel or a tortoise with a deer.

An animal's movements are related not only to its environment but also to its way of life.

Q1 Can you think why an animal needs to move from one place to another? Make a list of as many reasons as you can.

This chapter is concerned with movement in vertebrate animals. Even within this group there is great variety. Some methods are characteristic of particular animals but most use different methods of getting about at different times. For example, you may think of birds as flying animals but all have to come down at some time and must be able to walk or swim as well.

Whether animals move by swimming, running, or flying, there are some basic principles which apply to all of them. You can try to find what these are by using a simple trolley. The instructions for this are given in *figure 17*. Note particularly how the positions of trolley and person change during the experiment. To describe your observations there are some technical terms which you may find useful to make your explanation clearer. Two of these are *thrust* and *drag*.

Thrust is the action which an animal uses in pushing against its environment in order to move.
Drag is the force which the environment exerts to resist an animal's forward movement.

Figure 17
How to do the trolley experiment.

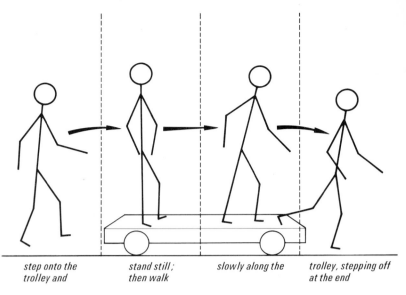

step onto the   stand still;   slowly along the   trolley, stepping off
trolley and   then walk   at the end

When you think you understand what is happening in the trolley experiment write an account including the words *thrust* and *drag*.

Other words which you may find useful are:

gravity   friction   support   reaction.

You may now see that movement is the outcome of a number of interacting forces which all have an effect on the final result.

2.2 Movement in land mammals

Three of the commonest ways in which mammals move on land are by walking, running, and jumping. In the trolley experiment you have been examining the effect of a human walking over a movable object. You may have noticed some of the actions which together are described as walking. Movement is brought about by the contraction of muscles which are attached to bones by tendons. The muscles pull on the bones which move because they are pivoted at a joint. (Joints are dealt with in more detail in *Living things and their environment*, Chapter 9, and in the Background reading at the end of this chapter.) The drawing of the human skeleton (*figure 18*) will help you to find the names of particular bones.

Ask one of your friends to walk across the floor. Observe carefully the way in which trunk and limbs are moved. Now look at the picture of athletes during a race (*figure 19*) and the drawing of one of the athletes (*figure 20*) which shows the main muscles in the trunk and limbs.

Figure 18
The human skeleton.

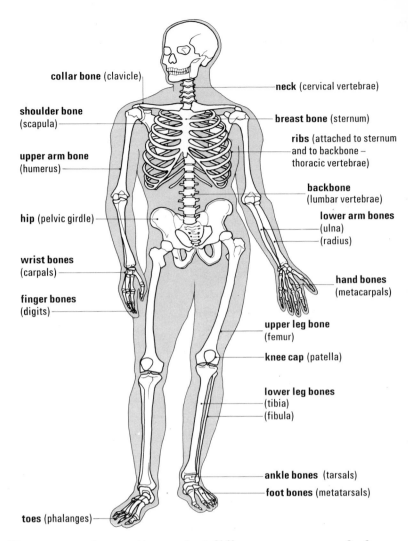

collar bone (clavicle)

neck (cervical vertebrae)

shoulder bone
(scapula)

breast bone (sternum)

ribs (attached to sternum
and to backbone –
thoracic vertebrae)

upper arm bone
(humerus)

backbone
(lumbar vertebrae)

lower arm bones
(ulna)
(radius)

hip (pelvic girdle)

wrist bones
(carpals)

hand bones
(metacarpals)

finger bones
(digits)

upper leg bone
(femur)

knee cap (patella)

lower leg bones
(tibia)
(fibula)

ankle bones (tarsals)

foot bones (metatarsals)

toes (phalanges)

Q1 From your observations what differences can you find between a walker and a runner?

Q2 Why does a runner move faster than a walker?

In walking and running as with all locomotion where limbs are used, once the limbs have thrust against the environment, they must regain their original position, relative to the trunk, so as to be ready to thrust again. This action can be described as striding. A *stride* is defined as the distance between the place where one foot touches the ground and where the *same* foot next touches the ground. If you have time and space you can try out some of the following suggestions.

Figure 19
A sprint race, in the 1968
Olympic Games in Mexico.
Photograph, E. D. Lacey.

Figure 20
The main muscles used by a runner.

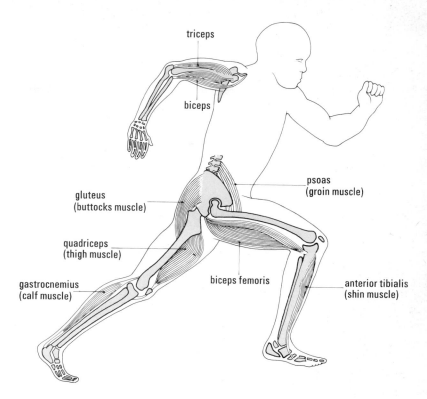

1 Find a method of measuring stride length.
2 Compare your own stride length *a* when you are walking with least effort, *b* when you are running at a pace suitable for a long distance race, and *c* when you are sprinting.
3 Measure your speed for all three activities.

   When calculating speed you will need to find the distance covered in a fixed time. Usually this is expressed as kilometres per hour or metres per second.

4 Measure length of *a* leg from hip joint to floor, *b* upper leg (femur), and *c* lower leg (from knee to floor).
5 Collect measurements from other people for all of points 1–4.

Q3 When you have collected all your results see if you can find any measurements which appear to be connected. For example can you see any relationship between speed and length of limb?

2.21 Fast-running mammals

So far all the work you have done has been concerned with human movement, but man is unusual amongst land mammals because he walks on two legs only and is erect.

Figure 21
Immediately after the start of a
greyhound race.
*Photograph, G. R. A.
Promotions Ltd.*

The greyhound in *figure 21* is much more characteristic
with its four legs, but if you compare it with the athletes in
figure 19 you will see many similar features. The fastest a
man can run is about 40 km per hour (25 m.p.h.) but a
greyhound can reach 65 km per hour (40 m.p.h.) in a race.
A cheetah can run really fast. At full stretch it moves at
about 110 km per hour (70 m.p.h.). *Figures 22a* and *b* show
the cheetah in action.

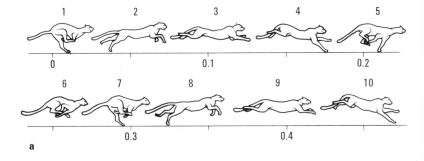

Figure 22
a A cheetah running fast.
b The fore and hind legs,
shoulders, hips, and backbones of
a cheetah running fast.

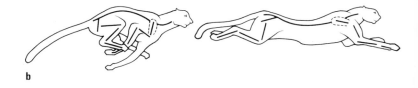

Living things in action

Q4 Make a list of all the things which you can see in the drawings in *figure 22* which may contribute to the cheetah's speed.

Fast animals need to take rapid strides but they may not be able to keep up a high speed for any length of time. This will depend on their endurance or the economic and efficient way they use their energy. A cheetah, for example, can only keep up its maximum speed for short bursts and its normal running speed is 45–65 km per hour. A giraffe can reach a speed of 55 km per hour if chased but it cannot sustain the effort for long because the blood supply to the brain is slow – and no wonder, for the blood has such a long way to go upwards from heart and lungs to the brain.

Fast-moving animals tend to have fairly small muscles since the bigger a muscle is, the slower the rate at which it contracts. The feet and lower limbs are light and can move rapidly. The muscles which work them are positioned high up on the limb. *Figure 23* shows the ligaments in a horse's foot. When the horse is galloping the hoof touches the ground very briefly. The ligament stretches, flexing the limb to absorb shock. Because of its elasticity it quickly springs back into position, straightening the lower joints rapidly. This kind of structure reduces the need for muscles.

Figure 23
The springing ligaments in a horse.

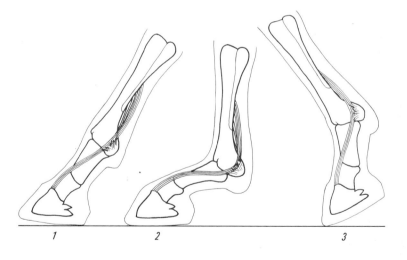

Animals which move slowly but powerfully show opposite tendencies to the fleet of foot. *Figure 24* shows comparable limbs from a sprinting mammal and a burrower. *Figure 25* shows two Olympic athletes, one a sprinter and one a wrestler. *Figure 26* shows two breeds of horses.

Q5 In each case explain, giving your evidence, which is the fast mover and which is the slower, more powerful, mover.

Figure 24
One of these mammals feeds by hunting over flat grassland; the other makes deep burrows and is a scavenger.

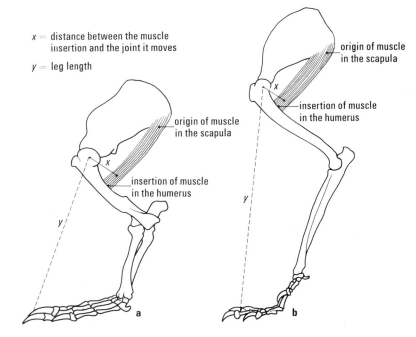

$x =$ distance between the muscle insertion and the joint it moves

$y =$ leg length

origin of muscle in the scapula

insertion of muscle in the humerus

origin of muscle in the scapula

insertion of muscle in the humerus

a

b

Figure 25
The profiles of two Olympic athletes.
From photographs in Tanner, J. M. (1964) The physique of the Olympic athlete, *Allen & Unwin.*

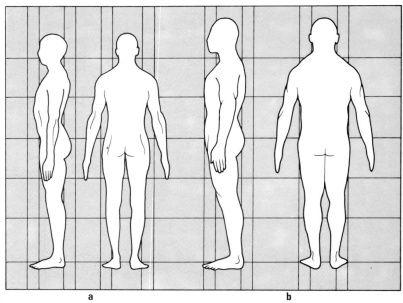

a

b

Living things in action

Figure 26
Two different breeds of horse.
Photographs; **a** *W. W. Rouch
& Co. Ltd;* **b** *Shire Horse Society.*

2.22 Animals which jump

Study the sequence of pictures in *figure 27a* and *b*. Although the frog makes a standing jump while the horse is making a running jump, there are a number of similarities; try to spot them.

Q6 What is the special function of the back legs?

Q7 What do the front legs do?

Q8 Which invertebrate animal shows similar body design and is thus enabled to jump?

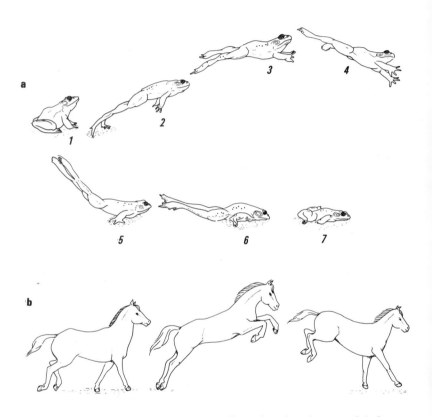

Figure 27
a The sequence of movements in a frog's leap.
b The sequence of movements as a horse jumps a fence.

Good jumpers tend to be small animals since a high body mass would require a large take-off thrust. If you look up the heights and distances to which animals can jump, you may see that small animals can jump relatively higher than large ones. Do you remember from Chapter 1 how leg size depends on body mass? Very large animals, with proportionally much larger legs, make poor jumpers.

Consider how flexible the legs shown in *figure 28* will be. If an elephant had to jump any great height, the shock on landing would be transmitted up the pillar-like column of legs and bones and girdle, breaking the bones or wrenching the animal's shoulder. It may be that landing from a jump is every bit as important as the initial take-off.

Figure 28
The hind legs of an elephant, a horse, and a cheetah.

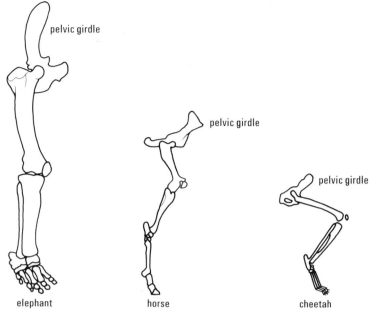

find the position of the heel bone in each animal

2.3 Flight of birds

You can learn a lot about flight by watching birds in action. Whether you live in the town or in the country you will see some birds in the air and such terms as 'gliding', 'flapping flight', and 'stalling' may mean something to you.

Q1   In *figure 29* (overleaf), which bird is illustrating each of these three flight activities?

If you can, watch rooks or gulls soaring high in the air on convection currents or thermals and see how a kestrel uses its wings and tail when it hovers. Next time you eat chicken, try to save all the wing bones and organize them as they would be in a living bird. The wing is a modified forelimb. The hand bones have been greatly reduced but there is a 'first finger' (*alula*) and two other fingers. Identify all these parts and compare the rest with the human skeleton (*figure 18*). There are three sorts of wing feathers and they vary in shape and stiffness. They are arranged in a very definite way and are attached to particular bones: primary flight feathers are attached to the finger bones, secondary flight feathers are attached to the ulna, while contour feathers cover the bird's wing. If you have a bird's wing, or, failing that, a photograph

Figure 29
Three gannets, each performing
different actions of flight.
Photographs, J. Barlee.

such as that of the owl's wing in *figure 30*, identify these
three types of feather. See if you can find how they compare
with the representative feather in *figure 31*.

If you have a pet bird, you could look at it to check your
ideas. You will see later how the shapes of these feathers
are related to the part they play in flight.

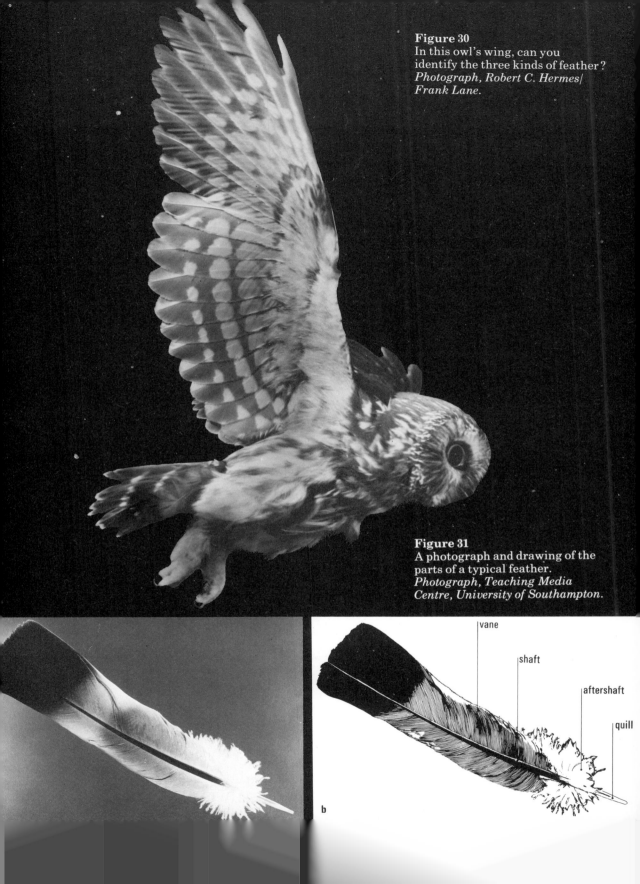

Figure 30
In this owl's wing, can you
identify the three kinds of feather?
*Photograph, Robert C. Hermes/
Frank Lane.*

Figure 31
A photograph and drawing of the
parts of a typical feather.
*Photograph, Teaching Media
Centre, University of Southampton.*

vane

shaft

aftershaft

quill

b

2.31 How a bird stays in the air

Flying is difficult because air gives little support and the bird's body tends to drop to the ground. To understand flight you have to realize that air presses in all directions – up, down, and sideways. Both the bird and the air are moving and, as you saw with the trolley, the final result is caused by the interaction of a number of forces. Those that affect flying are shown in *figure 32*.

Figure 32
A diagram to show the forces acting on a bird's wing.

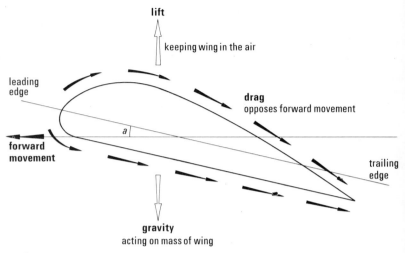

angle *a* is the wing's *angle of attack*

'Lift' can only happen if the pressure on the underside of the wing is greater than on the upper surface.

Apparatus can be designed to study the flow of liquid around different shapes. If you assume that air and liquid behave in a similar way, you will be able to gain some idea of how the position of a bird's wing alters the flow. You can work out how this could be done. One method is shown in *figure 33*, where coloured streams are produced from fine jets.

Figure 33
A method of finding how a bird's wing produces lift.

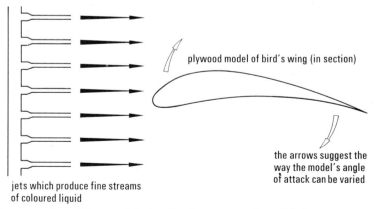

You may be able to think of other ways in which water currents could be marked. Whatever method you use, look carefully at the path of the marked streams of liquid;

places where they are drawn closer together indicate *high pressure*, places where the streams are drawn farther apart show *low pressure*. Look also for swirling and eddies (turbulence) some distance from the wing model.

Q2 Which particular wing position produces a great deal of turbulence?

Turbulence adds to the *drag* which opposes the wing's movement through the air. When it is very high, stalling results and the bird begins to fall. Birds do intentionally stall, not only for landing but also for quick manoeuvres.

You will have noticed that the alula bone has feathers attached to it and may have wondered if it is important. In 1963 a scientist called R. H. J. Brown set wings of freshly killed pigeons in a wind tunnel. In each, he sewed down the alula so that it could not move. By gradually increasing the wing's angle of attack (see *figure 32*) he reached a point where the feathers on the tilted wing became ruffled and in fact the wing was stalling. But on specimens with the alula free to move, tilting the wing to the same angle caused the alula to lift and the feathers did not ruffle. You will realize from this why the alula's movement is important. It is simple to modify the experiment shown in *figure 33* so that you can test this hypothesis. See if you can do it.

2.32 Powered flight

So far we have thought of the wing remaining stiff like the wings of an aeroplane. Although we can now understand what is meant by 'lift' and how stalling is prevented, we have not seen how 'drag' is overcome by the bird propelling itself forward. The bird is propelled by wings moved by enormous breast muscles which can make up a fifth to a half of the bird's total mass. High speed ciné photography, which can produce a film to be run in slow motion, gives a good idea of how the wing moves. *Figure 34* shows in 12 pictures a single flap of the slow flight of a pigeon taken from a film made by Brown. They have deliberately not been printed in the correct order.

The first picture is *D* and the last picture is *A*.
Put the letters in the order in which you think the pictures occurred in the film.

Look very carefully for identification clues. It may help you if, for each picture, you work out whether it shows:

	1	*2*
a	alula raised	*or* alula lowered
b	the wing moving backward	*or* the wing moving forward
c	the wing moving upward	*or* the wing moving downward

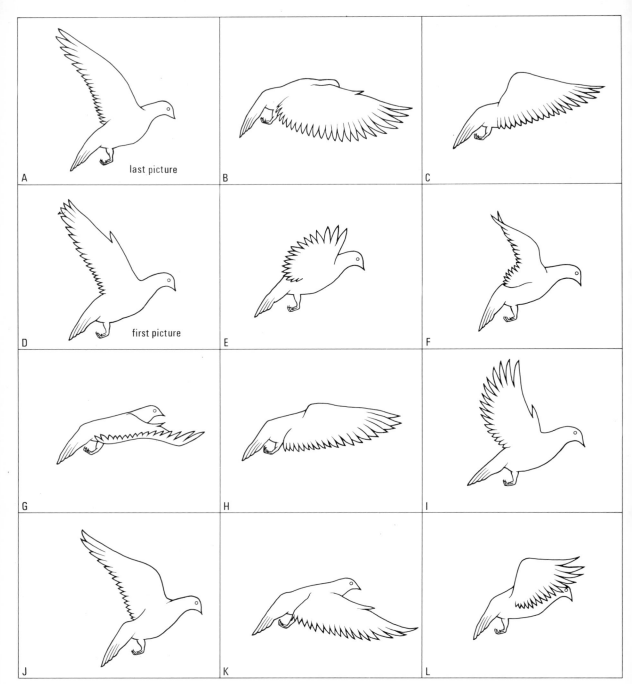

A
last picture

B

C

D
first picture

E

F

G

H

I

J

K

L

Figure 34
A pigeon in flight. The frames have been arranged in a random sequence. Note that the left wing is not shown in any frame.
Drawings based on photographs by Dr R. H. J. Brown. From Gray, J. (1953) How animals move, Cambridge University Press.

When you have decided on an order, check it with your teacher. If you have your own set of pictures, you can fix them in order in your own book or file and write below each one which of the alternatives in a, b, and c you chose.

2.33 How shape helps to overcome drag

You may have realized from the experiment tracing the paths of water streams (*figure 33*) that the shape as well as the angle of a bird's wing is important in affecting the flow of air round it. You could use the apparatus again and find out what would happen if a bird's wing was square in cross-section. A bird, however, is not composed of wings alone; it has a head, body, and feet, all of which will produce some resistance to air flow in flight. Here is a simple experiment (*figure 35*) which you can do to find if shape and wind resistance or drag are related.

Figure 35
An experiment to investigate the importance of a bird's overall shape.

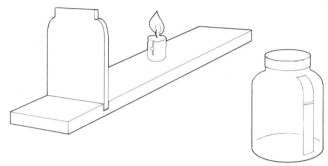

the cardboard shape is the exact size of the bottle, and fits into a groove in the board

1 Light the candle and stand it about 10 cm behind the groove in the board.
2 Stand the cardboard shape in the groove.
3 Blow the candle out by blowing at the front face of the cardboard shape.
4 Change the cardboard shape for the bottle, putting the middle of the bottle over the groove on the board.
5 Blow the candle out by blowing at the front of the bottle.

Q3 Note your results and explain why you think it was easier with one than the other.

Q4 Look at the photograph of the missel thrush (colour *plate 1*). Which is more like it in shape – the cardboard or the bottle?

Q5 Suggest how you think the shape of the head and rest of the body of a bird is adapted to overcome drag.

Q6 What do you think a bird does with its feet when in flight, and why?

From the work in this section you should now have some idea of how a bird is able to move through the air without having to depend on the support of its legs like mammals do when they move on land.

2.4 How fishes swim

Birds are highly adapted to efficient flight. Fishes show equally striking adaptations to swimming. Watch a fish swimming in an aquarium. Look at *figure 36* and identify the various fins on your animal.

Figure 36
Diagram of a typical fish. How is it adapted for swimming?

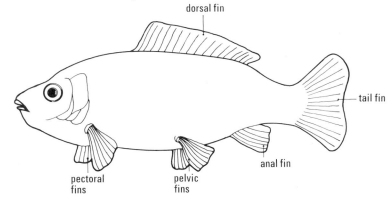

Q1   What part does the tail with its fin play in movement?

Q2   What movements do the other fins make and when do they make them?

Q3   Besides the fins what other parts of the fish are concerned with movement?

You will see from your observations that a fish is not only able to move forwards, but upwards, downwards, and even backwards. It can remain almost still and steady, roll from side to side, pitch by tilting from nose to tail, or move diagonally to and fro (yawing). The fish may make some movements of its own accord but others may be the result of its structure or of external forces such as water currents. A number of experiments are suggested which you can perform to find how a fish is able to move so efficiently in water. The results will supplement the ideas you had from watching the fish in the aquarium and you can use them to add to your explanations or correct them if necessary.

Shape and speed of movement
1   Obtain 9 pieces of Plasticine of identical mass (1 g or 2 g each).
2   Make 3 sets of 3 identical shapes. One set should resemble a fish but make the other sets as different as possible.
3   Put one ball of lead shot in the front end of each shape.
4   Drop a model into a tube as shown in *figure 37*, using a stopwatch to time how long it takes to fall down the tube from the top mark to the bottom mark.
5   Repeat your timing with each model so that you have three readings for each shape.

Figure 37
An apparatus for testing what the shape of a fish has to do with its movement in water.

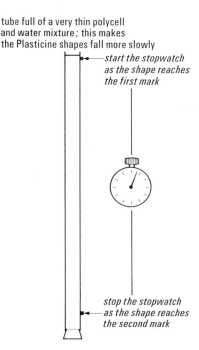

tube full of a very thin polycell and water mixture; this makes the Plasticine shapes fall more slowly

start the stopwatch as the shape reaches the first mark

stop the stopwatch as the shape reaches the second mark

Q4 From your results, which shape moved fastest through the liquid? Explain why.

Keeping steady
Design and make a simple model which you could use to find out how a fish is able to remain steady in the water. Remember, a number of fins may be concerned to cut down excessive movement. If you design your model well, you will be able to test the effect of each type of fin in turn, when the fish is moving in a variety of ways.

Q5 Which fins are most important in preventing the fish from *a* pitching, *b* rolling, and *c* yawing?

Look at the fish in the aquarium and check your answers.

Q6 Are there any actions of the fins which your model gives no evidence about?

Moving forward
Try making a model fish with a flexible tail which will help you to investigate the part played by the tail in forward movement. An apparatus of the kind shown in *figure 33* (page 30) would be very useful. Your results will be clearer if you make sure that the lower edge of the tail is in contact with the base of the apparatus. Record what you see and explain how the tail is used to propel the fish forward.

Thrusting against the water

By now, from observation and experiment, you will probably have a good idea of which part of a fish thrusts against its environment. You will be supplied with a fish, a scalpel or knife, scissors, and forceps.

1   Decide where the fish's thrusting muscles must be.
2   Slice off the top of the fish from head to tail, just above the backbone.
3   When you have exposed the muscles, try bending the fish as it would bend when thrusting against the water.
4   Look very carefully to see if you notice any changes in the muscles as you bend the fish.
5   Make a record of your investigation.

2.5 Skeletons and thrust

Figure 38 shows photographs of three skeletons. It is obvious what type of animals they have come from and you know how these animals move.

Figure 38
Three skeletons.
Photographs, Trustees of the British Museum (Natural History).

Q1   What part of each animal is mainly responsible for thrusting that animal forward?

Q2   What evidence can you see in these photographs to support your answers to question 1?

2.6 Movement in animals and plants

In this chapter you have been looking at movement in three particular groups of animals – land mammals, birds, and fish. You have seen how their shape, the way in which their limbs function, and their environment are all related. More attention has been concentrated on those types of animal which are rapid movers but you were reminded at the start of the work that there is an enormous range of performance, even within individual groups. In some animals other activities are more important than moving fast.

You may have noticed that the whole chapter has been confined to animals. This is not because plants are incapable of movement. Here, all the examples have been limited to those organisms which move from place to place. Some plants are also able to do this but most stay rooted in one situation. However, you should remember that even these can move their various parts.

Background reading

Joints

A joint is the place where two or more bones meet. Some joints are fixed – the bones cannot move relative to one another; but those joints which allow movement are much more interesting. You could investigate your own body to see what sorts of movement are possible at different movable joints. The X-ray photographs in *figure 39* overleaf illustrate four different parts of the body and you could use these as extra evidence.

How movable joints work
Two main requirements of a movable joint are that it should allow movement with little resistance – friction should be low – yet the bones should be held together securely (the joints should not dislocate). In *figure 40a* and *b* you can see some of the structural features which make this possible. The synovial membrane produces a fluid which lubricates the joint; your knee contains about 0.5 cm^3 of synovial fluid. You should be able to work out the function of the articular cartilage and the ligaments.

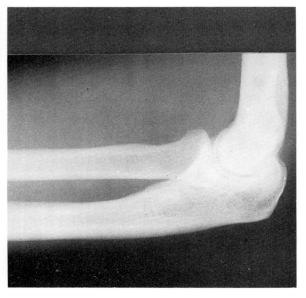

a Elbow

b Ankle

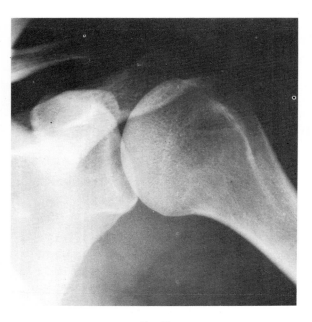

c Shoulder

d Neck

Figure 39
X-rays of four body joints.
*Photographs, Teaching Media
Centre, University of Southampton.*

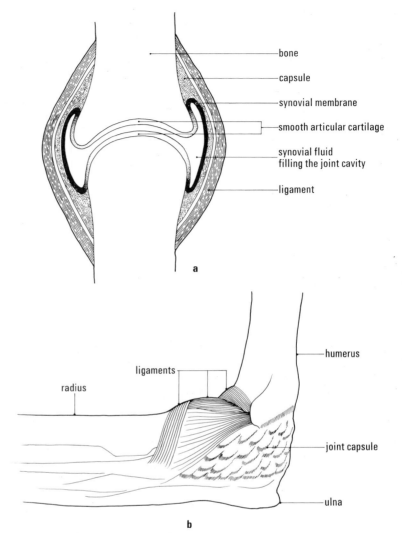

Figure 40
a The structure of a typical movable joint, as seen in longitudinal section.
b The left elbow, showing ligaments and joint capsule.
After Davies, D. V., and Davies F. (Eds) (1962) Gray's anatomy, 33rd edition, Longman.

bone
capsule
synovial membrane
smooth articular cartilage
synovial fluid filling the joint cavity
ligament

a

ligaments
radius
humerus
joint capsule
ulna

b

New joints for old
Ageing joints sometimes give trouble. In osteoarthritis, a disabling condition found in the elderly, the articular cartilage becomes rough. The cause of this is not fully known – perhaps certain foods are not getting through to the cartilage, or perhaps certain strains in movement bring the condition about. As osteoarthritis advances, the cartilage may wear away and, if bony outgrowths form, the joint will become stiff and painful.

A recently developed operation involves the complete removal of the faulty joint. A replacement hip, for instance, consists of a metal ball with a shaft inserted in the femur; the ball fits into a plastic cup fitted in the pelvis. Look at the X-rays in *figure 41*. The completed replacement is obvious enough. Try to decide which photograph shows a diseased hip and which a healthy one. Implant surgery such as this shows how modern medicine and engineering design come together.

Figure 41
X-rays of three hip joints.
Photographs, Teaching Media Centre, University of Southampton.

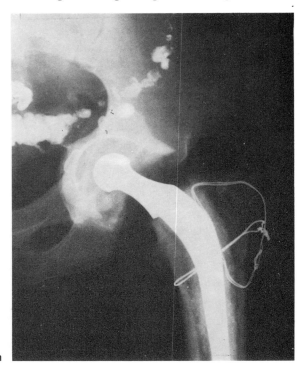

a

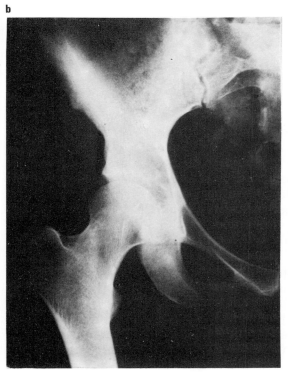

b

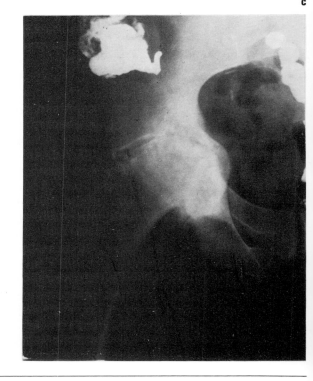

c

Living things in action

3

Maintaining a steady body temperature

3.1 Keeping the temperature constant

In this book we are finding out about living things in action. To be capable of action, a living thing requires energy. Energy, stored up in food, which has been eaten, is released and made available by chemical processes in the body. These processes only work rapidly over a certain range of temperature, of about 25 ° to 40 °C, with most working best at around 30 ° to 35 °C. If the temperature is too low, the chemical processes slow down and the animal becomes sluggish or even stops functioning altogether. Similarly, if the temperature is too high, the processes will stop.

Figure 42
Maintaining a steady body temperature despite the weather. (See also colour *plate 2*.) *Drawing by Mary Thomas.*

Tables 2 and *3* give information about some members of two groups of animals, mammals and birds. Study the tables carefully.

Name	Mass (kg)	Normal temperature and range (°C)	Climatic conditions of habitat (°C)
bear (brown)	550	38.0 ± 1.0	cold continental climate: −7° to 21°
bear (Polar)	850	37.5 ± 0.4	Polar: −34° to −1°
cat (domestic)	2.5	38.6 ± 1.3	temperate: 4° to 15°
camel	450	37.5 ± 0.5	tropical: going above 43°
eland (antelope)	900	38.8 ± 4.0	tropical: going above 43°
elephant	6600	36.2 ± 0.5	tropical: going above 43°
fox (red)	4.6	38.8 ± 1.3	temperate: 4° to 15°
horse	650	37.7 ± 0.5	widely distributed
man	70	36.9 ± 0.7	widely distributed
mouse (harvest)	0.03	39.3 ± 1.3	temperate: 4° to 15°
rhinoceros	4064	37.6 ± 0.2	tropical: going above 43°
shrew (European)	0.02	35.7 ± 1.2	temperate: 4° to 15°
whale	102180	35.7 ± 0.1	seas: Polar and temperate; not below 4°

Table 2
Data related to the temperature of mammals.

Name	Normal temperature and range (°C)	Climatic conditions of habitat (°C)
duck (mallard)	43.1 ± 0.3	temperate: 4° to 15°
gull (herring)	42.3 ± 0.7	temperate: 4° to 15°
ostrich	39.2 ± 0.7	tropical: going above 43°
penguin (little)	39.0 ± 0.2	cool temperate: 0° to 10°
skua gull	40.0 ± 0.5	Polar: −34° to −1°
thrush	40.0 ± 1.7	temperate: 4° to 15°
wren	41.0 ± 1.0	temperate: 4° to 15°

Table 3
Data related to the temperature of birds.

Q1   Can you see any similarities or differences between the two groups of animals where their temperatures are concerned? Make a list of any which you find.

Both birds and mammals can be described as *homoiothermic* because their body temperature is more or less steady.

3.2 Size and body temperature

The heat energy required by living things is produced by chemical processes which take place in all living cells (see Chapter 6). The amount of heat energy produced will therefore be related to body size. In *table 2*, the mass of the animal is given, not the volume. This is because the figures for mass are more readily available, but the mass usually gives some guide to the relative size of an animal. From *table 2*, you can see that the mass of an elephant is about 300 000 times as great as the mass of a shrew. Therefore, an elephant is considerably larger than a shrew and its capacity to generate heat will be correspondingly greater.

An animal's temperature will result from the balance between heat generated and heat lost. You can find, by means of an experiment, whether heat loss is affected by size.

1 Take two tin cans, a big one and a small one, and rinse each one out with very hot water.

2 Place them on the stands provided so that they are as completely surrounded by air as possible.

3 Find a method of filling the cans so that each of them will be as nearly as possible at the same temperature, at the start of your readings. Then fill the cans with nearly boiling water, *very carefully*.

4 Start your recording from the same temperature in each can and continue at 5-minute intervals for 30 minutes.

5 Draw a graph of your results. Plot temperature on the vertical axis and time on the horizontal axis. Compare the curves for the two cans.

If these two cans were homoiothermic animals, as soon as they lost heat energy they would have to generate more heat energy to keep the body temperature steady.

Q1 Which do you think would have the greater difficulty in maintaining its body temperature, a shrew or an elephant? Explain why.

Konrad Lorenz writes:

'All shrews are particularly difficult to keep; this is not because . . . they are hard to tame but because the rate at which food is used by these smallest of mammals is so fast that they will die of hunger within two or three hours if the food supply fails.'

Peter Crowcroft has shown from experiments that every day a shrew eats high protein food equal to 75 per cent of its body mass.

Q2 Why do you think a shrew has to eat such a large amount of food in relation to its size?

3.3 Size and surface area

Heat energy is produced in the body of an animal and is, therefore, related to its volume. It is the surface of the animal which comes into contact with the external world. Therefore, the area of the animal's surface greatly influences transfer of heat energy to or from its surroundings.

The next relationship to be considered is between size and surface area. For this work you can start by using simple cube models as in Chapter 1, because they are easier to understand than animals which have complicated shapes.

Length of side (cm)	Volume (cm³)	Surface area (cm²)
1	1	6
2		
3		
4		
5		
6		
7		
8		

Table 4

Using a series of cube models of increasing size, work out what happens to surface area in relation to volume. Use small 1 cm cubes as building units, if you have been provided with them, or draw models, or work out the answers by calculation. If one of your small cubes has a side 1 cm long, then

its volume will be 1 cm³
its surface area will be $6 \times 1 \text{ cm}^2 = 6 \text{ cm}^2$

because each face is 1 cm² and it has 6 faces exposed at the surface. Build cube models of increasing size and record the volume and total surface area for each one. Continue until you have a cube model with a side at least 8 cm long. The volume will be equal to the number of small 1 cm cubes you use and the total surface area will be equal to the number of 1 cm cube faces exposed at the surface. Make a table of your results as shown in *table 4*.

Q1 Which is larger, the number under the heading 'Volume' or the number under the heading 'Surface area',
a at the start of your series, with a small model,
b at the end of your series, with a large model?

Q2 Which of the following statements do you agree with? As the size of the model increases:
a Its surface area becomes larger in relation to its volume.
b Its surface area becomes smaller in relation to its volume.
c Its surface area remains the same in relation to its volume.

Mammal	Mass (g)	Area of skin (cm²)
48-hour-old gerbil	3	
adult female gerbil	74	
4-day-old hamster	67	139
adult female hamster	681	616
adult mole	76	
adult male mouse	29	
adult male rat	410	405
adult grey squirrel	650	444
adult male pigmy shrew	5	

Table 5

Figure 43 shows tracings of the skins of five mammals.

Table 5 gives data about the mass of these and other mammals and of the skin areas of the other mammals.

Measure the areas of the skins in *figure 43* and add them to *table 5* but arrange the mammals in increasing order of mass. Look carefully at these columns of figures and compare mass and surface area for each animal. Try comparing surface area with mass in the same way as you compared surface area with volume when working with cubes.

Q3 What happens to the surface area in relation to mass as the mammals get bigger?

Consider the examples given in *table 6*.

Mammal	Mass (g)	Surface area (cm²)
baby gerbil	3	
adult gerbil	74	
baby hamster	67	139
adult hamster	681	616

Table 6
What is the ratio of the skin area to mass, in each of the cases?

Q4 Is your answer to question 3 still true if you compare mass and surface area of a baby and an adult of the same mammal?

3.4 Four penguins: a review problem

You will now have some ideas about the effect of size on temperature control in homoiothermic animals. Try to apply your knowledge to the penguins shown in *figure 44a*.

Q1 Each of the four penguins comes from one of the four places marked X on the map in *figure 44b*. Suggest which penguin lives in which place and explain why you have chosen each site.

Figure 44
Each of the four types of penguins in **a** comes from one of the places marked X in **b**.
Based on Hvass, H. (1961) Birds of the world, *Methuen.*

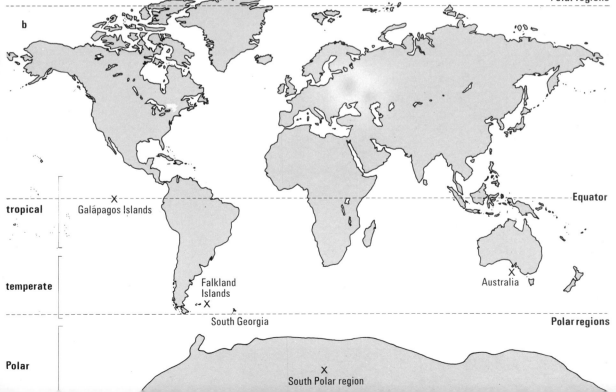

A
Height 48 cm

B
Height 38 cm

C
Height 102 cm

D
Height 68 cm

b

Polar regions

tropical — Galápagos Islands

Equator

temperate — Falkland Islands, Australia

South Georgia

Polar regions

Polar — South Polar region

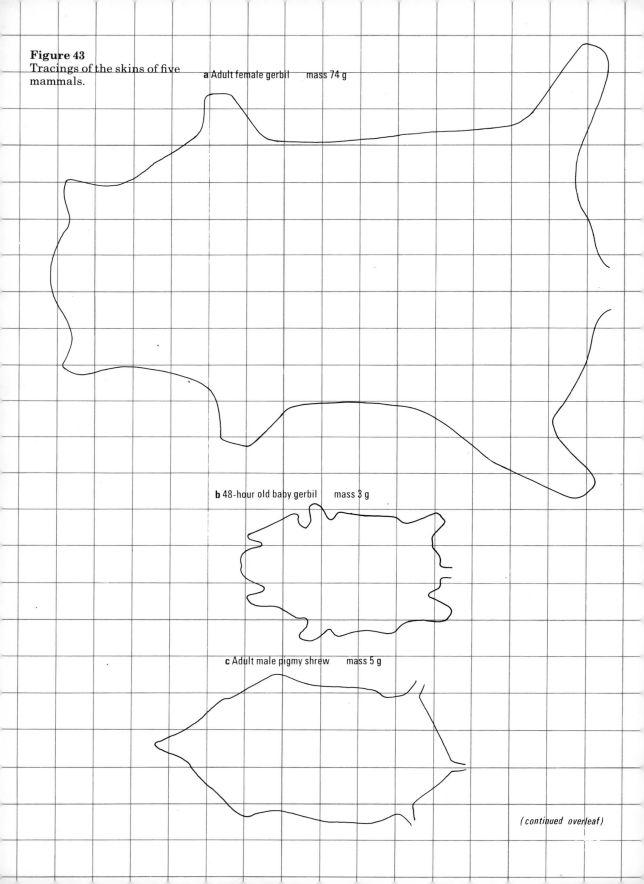

Figure 43
Tracings of the skins of five mammals.

a Adult female gerbil   mass 74 g

b 48-hour old baby gerbil   mass 3 g

c Adult male pigmy shrew   mass 5 g

(continued overleaf)

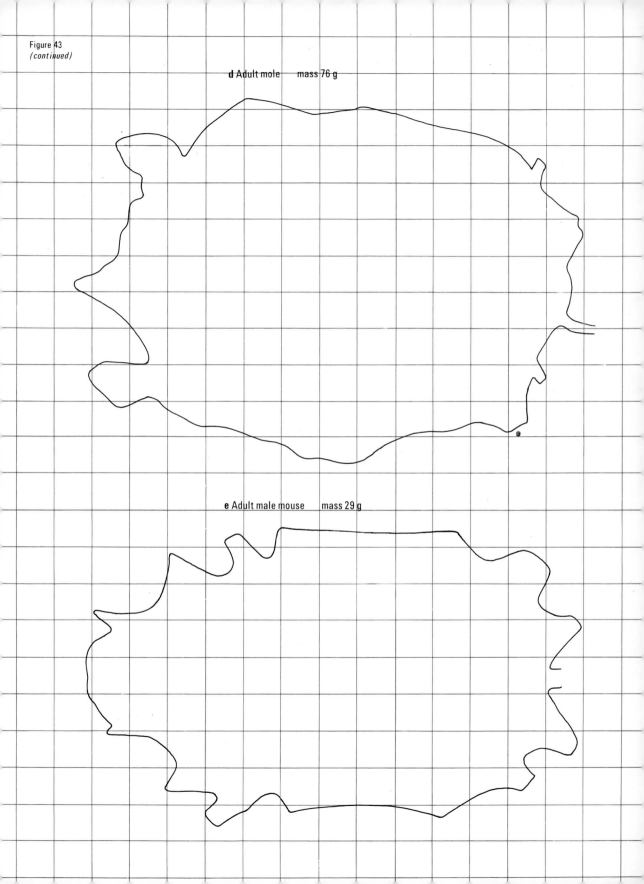

Figure 43
(continued)

d Adult mole   mass 76 g

e Adult male mouse   mass 29 g

Reindeer (*Rangifer tarandus*)
Scandinavia, Finland, Siberia,
Spitzbergen, western Greenland,
Arctic North America

Fallow deer (*Dama dama*)
native of southern Europe and
Asia Minor.

Roe deer (*Capreolus capreolus*)
Britain, southern Europe

Elk or moose (*Alces alces*)
Scandinavia, northern Russia,
North America

length from snout to base of tail

elk 198 cm

elk 280 cm

reindeer 178 cm

fallow 153 cm

reindeer 108 cm

fallow 89 cm
roe 74 cm

roe 102 cm

shoulder heights

Figure 45 (*opposite*)
Four kinds of deer and where they
are found.
Based on Hvass, H. (1961)
Mammals of the world, *Methuen.*

Living things in action

3.5 Mammals from different areas of the world

You can now consider how mammals maintain their temperature despite environmental conditions. *Figure 45* illustrates four species of deer, *figure 46,* two closely related cattle species, *figure 47,* three foxes, and *figure 48,* three bears. Within each group there should be an obvious family resemblance so that body shape is roughly similar in each. Study the *differences* between the members of a single group and see how far these are related to the animal's distribution – in polar, temperate, or tropical regions, for instance.

This is a rather difficult exercise, because more than one factor may be important; for instance, body size matters but may not be the reason for the whole story. Do not be disheartened by complexities – biology is full of processes which depend on interacting factors. If you apply the principles of size and surface area (plus a little common sense), you should go a good way towards finding possible solutions.

Once you have thought out some ideas which seem to solve the problem, in other words, once you have framed a hypothesis, you could devise an experimental test. Modifying the tin can procedure in section 3.2 may be one possibility.

Figure 46 (*right*)
The yak and the domestic buffalo and where they are found.
Based on Hvass, H. (1961)
Mammals of the world, *Methuen.*

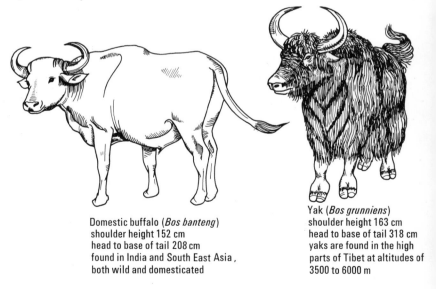

Domestic buffalo (*Bos banteng*)
shoulder height 152 cm
head to base of tail 208 cm
found in India and South East Asia,
both wild and domesticated

Yak (*Bos grunniens*)
shoulder height 163 cm
head to base of tail 318 cm
yaks are found in the high
parts of Tibet at altitudes of
3500 to 6000 m

Figure 47
Foxes and where they are found.
a *and* **b** *are based on Hvass, H.*
(1961) Mammals of the world,
Methuen; c is based on Dorst, J.,
and Dandelot, P. (1970) A field
guide to the larger mammals of
Africa, Collins.

a
Arctic fox (*Alopex lagopus*)
shoulder height 33 cm
length without tail 60 cm
inhabits Polar regions of the
Northern Hemisphere;
especially common all over
Greenland

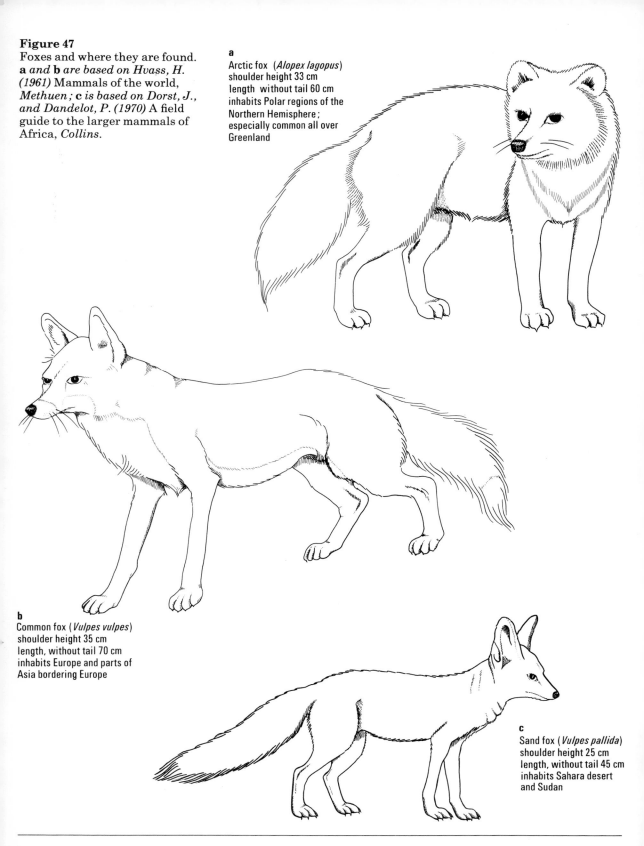

b
Common fox (*Vulpes vulpes*)
shoulder height 35 cm
length, without tail 70 cm
inhabits Europe and parts of
Asia bordering Europe

c
Sand fox (*Vulpes pallida*)
shoulder height 25 cm
length, without tail 45 cm
inhabits Sahara desert
and Sudan

Figure 48
Three kinds of bear and where
they come from.
Drawings by Mary Thomas,
based on Grassé, P. P. (ed.) (1955)
Traité de Zoologie, *Volume XVII,*
Masson et Cie, Paris.

Polar bear (*Thalarctos maritimus*)
shoulder height 140 cm
length, without tail 250 cm
found on Arctic coasts of
America and Greenland

Brown bear (*Ursus arctos*)
shoulder height 120 cm
length, without tail 208 cm
found in Norway, Sweden,
Finland, Russia

Malay bear (*Helarctos malayanus*)
shoulder height 68 cm
length, without tail 137 cm
found in South east Asia and
East Indies

3.6 A variety of problems

Each of the following sections is devoted to a single problem about temperature control. In trying to solve each, you could follow a basic pattern as you did in section 3.5.

1   Study all the information provided, both in words and illustrations.
2   Decide what factor or factors may be important and frame a hypothesis which might explain the problem.
3   Devise an experiment to test the hypothesis.
4   Check with your teacher that the experiment is a reliable one and that apparatus is available for it.
5   Carry out the experiment and see if the results confirm your first hypothesis. You may have to modify the hypothesis in the light of your experimental results.
6   Even if the experiment confirms your hypothesis, consider its limitations – for instance, how accurately or reliably were results taken, how far does the experiment oversimplify the problem, etc?

Problem A
When we are very hot, we sweat and it seems to cool us. When a dog is hot, it pants with its large wet tongue hanging out. Can you give an explanation for these observations? Can you produce any experimental evidence in support of your explanation?

Problem B
Penguins huddle together during the Antarctic winter, changing places with one another from time to time. During the winter, honey bees do the same in their hives. Sheep on upland pastures also stand close together in the cold. Why do they do this?

Design an experiment which will match the picture these descriptions give and which will act as a check on your explanation.

Problem C
Figure 49 shows drawings of robins taken from photographs made in summer and in winter.

Figure 49
Robins at different seasons.

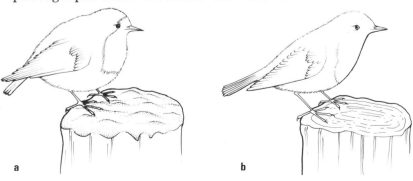

a   b

Living things in action

From the appearance of the birds, decide which was taken when. What differences do you notice in their appearances? How might these differences be important to the two birds? Plan an experiment to check your ideas.

Problem D
In Ethiopia the tribes living in the mountain regions are much more stockily built than the tall, slight plainsmen. One climatic difference is that the mountains are colder than the plains. Similar differences in body shape may be seen in colour *plates 2a* and *b*.

Suggest how body shape may affect temperature control. Carry out an experiment to test your ideas.

Problem E
When the tropical day is at its hottest, elephants will often stand in the shade with their huge ears spread out. What does this do? Can you prove what you say?

Problem F
'My wife caught an adult shrew on a path through thick grass late in summer. . . . It was very weak. This shrew was found after a thunderstorm, and a heavy downpour which lasted over an hour must have prevented it from finding food as well as soaking its fur, causing excessive heat loss.'
(From P. Crowcroft, The life of the shrew.*)*

Design an experiment to check just how much faster a furry object loses heat when its fur is wet through. Can you explain why wetting fur affects its insulating power?

Problem G
Twentieth-century man is adapted to his environment, but he also *adapts his environment* to suit his needs. What examples can you think of which support this idea?

Background reading

Hypothermia

All the work in this chapter has been concerned with animals which maintain a more or less constant temperature. These homoiothermic animals are sometimes loosely called warm-blooded. Poikilothermic animals such as insects and reptiles cannot regulate their temperature in the way that mammals can. It fluctuates according to their environment. These animals are often called cold-blooded but this is a misleading term because, if their environment is warm, then their blood will also be warm.

In homoiothermic animals the temperature is regulated very carefully. Should the body temperature fall below an acceptable level, a thermostatic centre in the brain sets off several reactions which bring the temperature back to normal. Here are some of the reactions.

1   Certain parts of the body, such as the liver, use up food more quickly and so generate more heat.

2   In *shivering*, the muscular activity liberates heat.

3   There is a reduction in the amount of blood flowing through the skin, hence a slower loss of heat into the air. This is why a person's skin looks pale when he is cold.

4   The hairs in the skin stand upright. In man this mechanism is of no value, merely producing goose pimples, but in most furry animals the insulating layer of fur is greatly thickened.

You should be able to think of other activities, some of which may be done consciously, that help to keep the body warm. Try to work out how we keep cool under hot conditions.

Sometimes the mechanisms listed are insufficient to fight off the cold; if body temperature falls below normal then an animal is said to be *hypothermic*. A hypothermic animal does not really function in quite the same way as a cold-blooded animal. Hypothermia can seriously disturb a normally warm-blooded animal and may lead to death, but cold-blooded animals do not have these difficulties when their temperature changes. Experiments show that in any animal, the chemical reactions of the body, the functioning of the body, and the animal's activity all become two or three times slower for each 10 °C reduction in body temperature. Thus a cold-blooded animal, such as an insect, a frog, or a crocodile, just moves at a slower speed when the temperature is lowered. In warm-blooded animals, cooling of the body produces, first, mental confusion and then, at 27 °C, complete anaesthesia. At about this same temperature shivering stops so that the body no longer resists the cold. Below 28 °C or so, breathing slows down and eventually stops at about 25 °C. In humans, breathing would stop at a lower temperature, but death occurs at about 25 °C because the heart stops beating.

Heat loss and body cooling are much faster in water than in air. Most people would die after 30 minutes in ice-cold water, and it is very unlikely that anyone could survive more than 90 minutes. It is probable that the cold will cause mental confusion, then anaesthesia, so that people may drown even though normally they are good swimmers:

people in life jackets die when their temperature falls to 25 °C. At 11.40 p.m. on 14 April 1912, the *Titanic* struck an iceberg and began to sink into an ice-cold sea. Shortly before dawn the *Carpathia* arrived and rescued the 712 people who had managed to get into lifeboats; but all 1489 people still in the water were dead, even though most wore lifebelts to keep them afloat. It is clear that cold was the main cause of death. It is now thought that cold is also one of the main causes of death in people lost around the coast of Britain; people who float out to sea will become hypothermic and die unless they are rescued in time.

Hypothermia is a serious problem on land too, and not just amongst Arctic explorers and mountaineers. In Britain, in the winter months, the cold probably affects very many people, especially the very young and the very old. The ability to regulate body temperature is not fully developed in babies, but we do not really know why the elderly become hypothermic; it may be due to a combination of many reasons. It is thought that the temperature-regulating mechanisms may lose sensitivity with age; many old people do not feel the cold and can become dangerously hypothermic without realizing it. Other reasons for hypothermia, of course, include inadequate heating, clothing, and diet, and general lack of fitness in the elderly.

Hibernating animals are different from other warm-blooded creatures. They are specially adapted to withstand body temperatures as low as 1–2 °C. Hibernation allows an animal to live very economically on its bodily food reserves at a time when natural food is scarce. During hibernation the animal is very still and inactive so that energy is only required for the essential bodily functions; but these essential functions, like all other processes, are slowed down by the low temperature and therefore consume less energy. The great reduction in the body's needs is seen in the change in heart rate; for example, the heart of a hedgehog normally beats three times a second, but during hibernation it beats only once in ten seconds. The hibernating hedgehog is not really a cold-blooded animal since it is in a sleep-like state, nor is it like a hypothermic animal since it can quickly generate the heat to rewarm itself and become active again if disturbed or in danger.

The scientist and the doctor have studied all of these conditions of low temperatures, and use their knowledge for man's benefit. Hypothermia is now sometimes used deliberately during complicated surgical operations. It was first used for heart surgery in 1953, by two American surgeons, Lewis and Tauffic. In those days the patients were made hypothermic by placing them in a bath of cold water. It usually took about an hour for the temperature to fall to about 30°C. A much more efficient method is now used instead. The surgeon connects an artery and a vein of the patient to a large coiled tube that is immersed in cold water. The blood flows from the artery to the tube and then back through the vein into the body. The blood is cooled as it passes through the coiled tube, and then the cold blood very quickly cools the body. It is like central heating – but designed to lower the body temperature, not to raise it. It is important to realize that the brain's thermostat must be switched off before hypothermia can be produced. This is done by anaesthetizing the person. One of the reasons for using hypothermia in an operation is that the surgeon is able to stop the blood flow and perform an operation without the difficulties normally caused by bleeding. During hypothermia the body needs less oxygen and less food materials; and so all parts of the body can therefore survive longer without blood flowing through them. For example, at normal temperatures the brain is permanently damaged if the blood flow stops for three minutes. At 28°C the brain is unharmed when the blood flow is stopped for ten minutes. Another reason for using hypothermia is that in heart operations, the cold stops the heart beating so that the surgeon can more easily repair it. Once the operation is over, the body is rewarmed and the heart is easily restarted.

4

Living and breathing

In most people's minds the process of breathing is closely connected with life itself. Indeed one can often tell if a person is dead or alive by testing whether or not he is breathing. It would therefore seem likely that there is a purpose in breathing. One hypothesis might be that we use, or in some way alter, the air going into our bodies before we pass it out again. You can test this simply.

1   Find out whether the air you breathe out (exhaled air) will support combustion for a longer or a shorter time than the air you breathe in (inhaled or atmospheric air). *Figure 50* shows a simple apparatus for collecting small samples of exhaled air, or even air that has been breathed in and out several times. The diagram also shows how samples of air can be tested with a lighted candle.

2   You can design the details of the experiment yourself. You should bear in mind that the two jars, A and B, although they contain different kinds of air, should be similar in every other way.

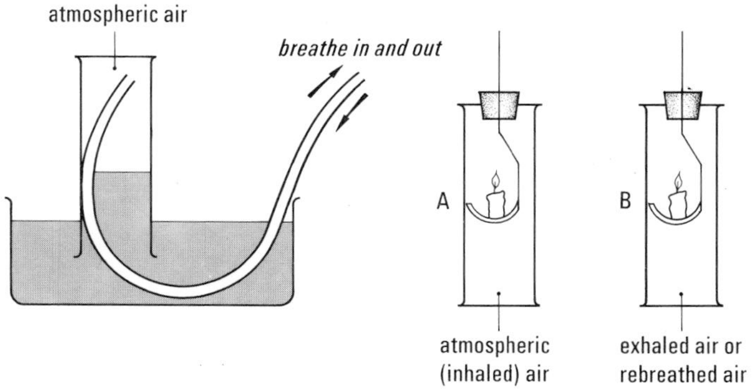

atmospheric air

breathe in and out

A   B

atmospheric
(inhaled) air

exhaled air or
rebreathed air

Figure 50
One method of collecting samples
of exhaled air and testing them
with a candle flame.

Compare your results with those of others in the class and then try to answer the following question:

Q1   How can you explain any difference which you found between how long the candle burned in atmospheric air and how long it burned in exhaled (or rebreathed) air?

4.2 Analysing the air

The experiment in section 4.1 will probably have suggested that there are some differences between exhaled and inhaled air. It will not have indicated the nature of these differences, although in your answers to question 1 in section 4.1 you may have suggested what they could be. One of your next questions should be, 'In precisely what ways is the exhaled air different from inhaled air?' It would be a long job to find the complete answer, and we shall confine ourselves to measuring two substances in the air – carbon dioxide and oxygen. We shall assume that the rest of the air is nitrogen.

In analysing the air we find out the proportions of carbon dioxide and of oxygen by removing each in turn from a measured volume of air. Two reagents are used: potassium hydroxide solution which absorbs carbon dioxide; *and*
potassium pyrogallate solution which absorbs both oxygen and carbon dioxide.

So the air is treated with each of these solutions in turn. The contraction of the air after each absorption will indicate the proportions of carbon dioxide and oxygen in the air sample. Both the reagents are caustic (harmful, especially to the skin) so if any should get on the bench, your clothes, or your skin you should wash it away *immediately* with plenty of water and report the mishap to your teacher.

Figures 51, 52, and 53
Three alternative pieces of apparatus which can be used for gas analysis.

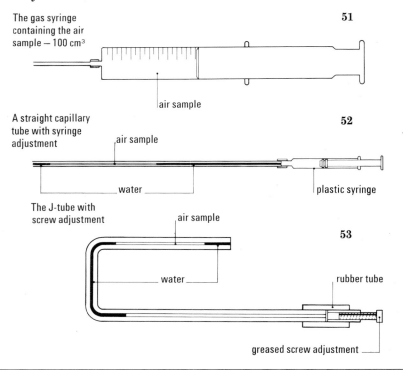

The gas syringe containing the air sample – 100 cm³

51

air sample

A straight capillary tube with syringe adjustment

52

air sample

water

plastic syringe

The J-tube with screw adjustment

53

air sample

water

rubber tube

greased screw adjustment

Living things in action

There are several ways of carrying out this analysis and you will be given the precise experimental details. Each of the methods involves the following stages:

1 Collect a sample of the air to be analysed in the syringe or tube to be used for the analysis. (*Figures 51* to *53.*)
2 Measure the volume of the air sample.
3 Introduce some potassium hydroxide solution to the sample and mix to absorb the carbon dioxide.
4 Measure the new volume of the sample. Any difference between this and the result in *2* will be equal to the amount of carbon dioxide in the original sample.
5 Introduce some potassium pyrogallate solution and mix to absorb the oxygen.
6 Measure the new volume. Any difference between this and the result in *4* will be equal to the amount of oxygen in the original sample.

Q1 Why is it important to absorb the carbon dioxide with potassium hydroxide solution first and then the oxygen with potassium pyrogallate solution, and not the other way round?

You can analyse both atmospheric and exhaled air by this method. Your samples must be at room temperature and pressure and some of the details of the procedure will be concerned with making sure that this is the case.

Q2 Calculate the percentages of carbon dioxide and oxygen in atmospheric and exhaled air.

Compare your results with those of others and calculate the mean (average) result for each percentage value.

Q3 What appears to have happened to the composition of the air while it has been in your body?

Q4 How variable are the class results for *a* atmospheric and *b* exhaled air?

Q5 What reasons can you suggest for these variations within the class?

Q6 What other differences do you think might exist between exhaled and atmospheric air?

4.3 Exchange of gases

So far we have been concerned with the breathing of humans. You will have seen that exhaled air differs in a number of ways from inhaled air. The alteration in the composition of the air takes place in our lungs. It is known as *gas exchange*. The question now arises as to whether other living organisms affect the air as you have done (and as you demonstrated in section 4.2). However, collecting samples of exhaled air from other living organisms (such as locusts or germinating seeds) might prove rather difficult.

Q1 Can you devise an experiment, involving gas analysis, to find out whether other living organisms carry out gas exchange?

Gas analysis, as you have seen, is quite a lengthy and complicated procedure. Your analysis will have shown that exhaled air contains more carbon dioxide than atmospheric air. A much simpler and equally sensitive test for detecting gas exchange involves testing the air around an organism for carbon dioxide. An increase in carbon dioxide concentration is fairly good evidence for gas exchange.

4.31 Detecting carbon dioxide production

For many years chemists have used lime water to show the presence of carbon dioxide. You are probably familiar with this test, in which the lime water becomes milky or cloudy in the presence of carbon dioxide. For our purposes a suitable test for carbon dioxide should be both *sensitive* and *specific*. By 'sensitive' we mean that the test must be able to detect a small amount of carbon dioxide. By 'specific' we mean that other gases should not affect the test in the same way as carbon dioxide. An alternative test to lime water uses bicarbonate/indicator solution. Bicarbonate/indicator solution actually detects changes in acidity and alkalinity (changes in pH).

1 Study the colour changes of bicarbonate/indicator solution by placing a little in each of three test-tubes, labelled A, B, and C. Add one drop of dilute acid to A and one drop of dilute alkali to B. Add nothing to C.
Carefully record the colours produced in each of the three tubes.

2 Take two tubes, D and E, containing bicarbonate/indicator solution, connected together in the way shown in *figure 54*. This ensures that the air breathed in passes through one tube, while the exhaled air passes through the other.

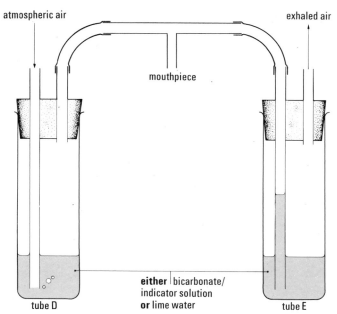

Figure 54
Apparatus for comparing the carbon dioxide content of atmospheric air and exhaled air.

atmospheric air

exhaled air

mouthpiece

either bicarbonate/
indicator solution
or lime water

tube D

tube E

Breathe *gently* in and out through the mouthpiece a few times. Note the colour of the indicator in the two tubes.

3   Repeat *2*, using lime water in the tubes instead of bicarbonate/indicator solution.
Now answer the following questions:

Q2   What is the effect of exhaled air on bicarbonate/indicator solution?

Q3   In comparing bicarbonate/indicator solution with lime water as tests for carbon dioxide, which is
a   the more sensitive?
b   the more specific?

Q4   Of the two tests, which do you think is the more suitable for detecting carbon dioxide production by small living organisms?

4.32 Carbon dioxide production in other organisms

You are now in a position to find out whether small living organisms carry out gas exchange. In using bicarbonate/indicator you must remember that very little carbon dioxide is needed for the colour change. You must therefore be careful not to breathe on the solution while you are setting up the experiment.

1   Rinse out several test-tubes with distilled water and then with bicarbonate/indicator solution. Then place 2–3 cm^3 of bicarbonate/indicator solution into each tube.

2   Push a perforated platform into each test-tube and then put some living material in all tubes but one (*figure 55*).

3 Check that the colour of the bicarbonate/indicator is the same in each tube at the start of the experiment.

4 Leave the tubes for at least half an hour, comparing the colour of the indicators from time to time. Note the final colour of the indicator in each test-tube.

Figure 55
Tubes for testing living things for the production of carbon dioxide.

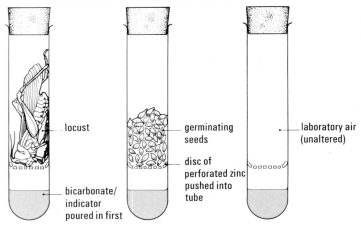

locust

germinating seeds

laboratory air (unaltered)

disc of perforated zinc pushed into tube

bicarbonate/ indicator poured in first

Now answer the following questions:

Q5 In one tube there was no living material.
a What name is given to an experiment like this, where one important part of the procedure is left out?
b What is the purpose of this tube?
c Why is a perforated platform placed in this tube?

Q6 Why is it so important to rinse out all the tubes before use?

Compare your results with those of other members of the class who have perhaps used a different selection of living organisms.

Q7 What does this experiment tell us about the production of carbon dioxide by living organisms?

Q8 What does this experiment tell us about the uptake of oxygen by living organisms?

Background reading

When oxygen is scarce

Climbing high mountains
If you carry a barometer up a hill you will see that atmospheric pressure falls as you climb. In other words, the higher you go, the less dense the air becomes. *Table 7* shows the readings a barometer would give at various altitudes.

Place	Approximate altitude (m)	Approximate atmospheric pressure (mm of mercury)
sea level	0	760
summit of Scafel	1000	675
Mexico Olympic Stadium	2000	595
summit of Mont Blanc	5000	400
summit of Everest	9000	230

Table 7

Since the air at the summit of Everest is less than three times as dense as it is at sea level and since the proportion of oxygen present is unchanged at about 21 per cent, it follows that a breath of air on Everest will contain less than a third as much oxygen as a breath at sea level. This oxygen shortage, of course, seriously impairs the process of respiration on which life depends.

If the pressurization of an aircraft failed at 8000 metres, or if a helicopter landed its passengers as high as this on a mountain, then the people concerned would soon become unconscious and would eventually die from lack of oxygen.

Yet mountaineers who climb slowly up a mountain may be able to live, work, and sleep at 8000 metres. As they ascend, their bodies adapt to the oxygen shortage by making changes which we refer to as *acclimatization*. Acclimatization involves not only breathing deeper and faster, but also certain other changes which enable the blood system to carry more oxygen.

This slow process of acclimatization has enabled mountaineers to climb many high peaks without artificial aids and encouraged a number of attempts on Everest itself during the first half of this century. It has been found, however, that acclimatization has its limits and that a prolonged stay at very high altitudes brings about a deterioration in the work capacity of the body. In the 1924 British Everest expedition, Colonel Norton, a man whose exceptionally low pulse rate indicated a very strong heart, acclimatized unusually well, yet by the time he had reached an altitude of about 8500 metres (about 500 metres below the summit) he was forced to take eight breaths between each step and to have a longer rest every few metres.

In the light of such experiences most Himalayan expeditions since the early 1920s have attempted to carry oxygen supplies for use at high altitudes. Yet this seemingly simple solution brings many problems, not least of which is transport. At a high work rate a man uses 5 dm^3 of oxygen per minute and at very high altitudes, climbers may also use 1 dm^3 per minute to help them sleep.

So, ideally, a mountaineer might require 5000 dm³ per day for the assault on the summit. Such an enormous volume would be far too cumbersome unless stored under pressure in cylinders. The cylinders themselves are heavy, even when designed specially for mountaineers, as they must be strong enough to withstand the pressure.

The British expedition which first climbed Everest in 1953 carried 60 aluminium alloy cylinders, each containing 800 dm³ and with a mass of more than 5 kg, as well as 100 steel cylinders, each containing 1400 dm³ and with a mass of nearly 10 kg. It is hardly surprising that the expedition needed 350 porters to carry their equipment to the mountain and then 40 Sherpas to ferry loads to the high camps. When Hillary and Tenzing set out for their final summit assault, their oxygen equipment had a mass of 14 kg each and only contained enough oxygen to allow 3 dm³ per minute. It is interesting to note, however, that they were able to do without extra oxygen for 10 minutes on the summit whilst Hillary took photographs.

Figure 56
Sir Edmund Hillary and Sherpa Tenzing, high up on Everest.
Photograph, The Mount Everest Foundation.

During the years that mountaineers have been using oxygen, scientists have been trying to improve the equipment. You have found, by analysing normal air and expired air, that the lungs extract only about one-quarter of the oxygen from the air they breathe in. The other three-quarters passes out again in the expired air. Obviously it is very wasteful to carry oxygen up a mountain, for a climber to breathe, if most of it will be lost when he breathes out again. A closed circuit system has been developed which conserves the oxygen lost by the original open circuit method. In a closed circuit system the expired air is circulated through a soda-lime canister, to remove carbon dioxide, and the air is then rebreathed after extra oxygen has been added.

Figure 57
A closed circuit breathing apparatus.

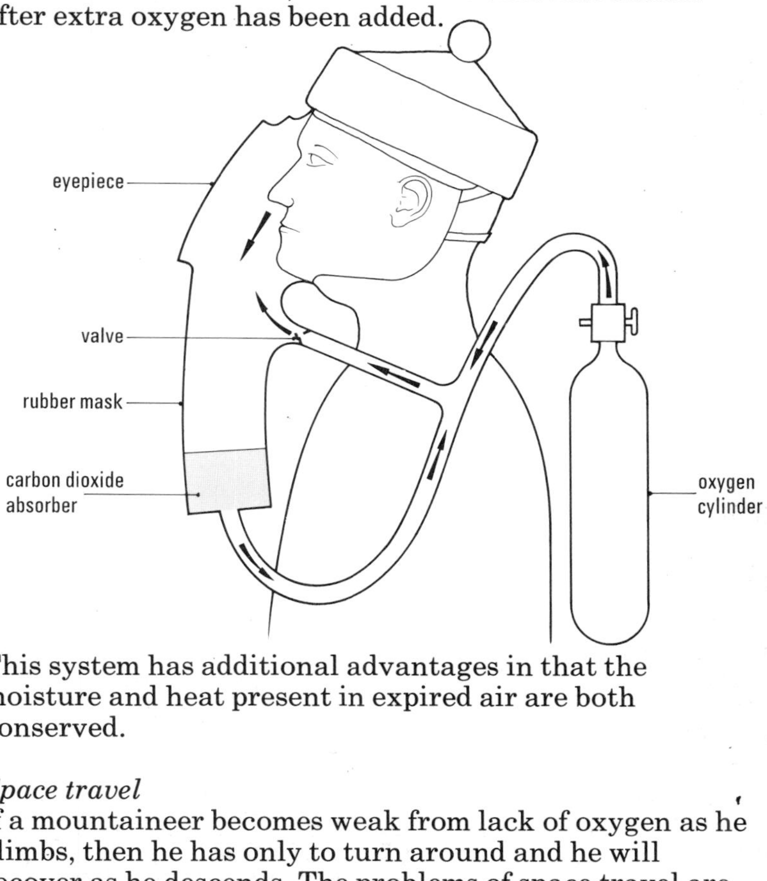

eyepiece

valve

rubber mask

carbon dioxide absorber

oxygen cylinder

This system has additional advantages in that the moisture and heat present in expired air are both conserved.

Space travel
If a mountaineer becomes weak from lack of oxygen as he climbs, then he has only to turn around and he will recover as he descends. The problems of space travel are less easily avoided since there is no air and therefore no oxygen in space. The space traveller depends entirely on the oxygen initially present in the cabin of his spacecraft and on whatever extra supply he has stored on board. The crew's oxygen requirement for the whole journey must be calculated accurately so that enough oxygen is provided.

Manned space exploration has been limited so far to the nearer parts of our solar system. One of the reasons for this is the difficulty of storing and lifting the vast amount of oxygen which would be required for a journey to the more distant bodies. We can speculate on ways of using simple plants to solve the oxygen problems in space. Brightly illuminated tanks of algae would photosynthesize and in doing so would evolve oxygen as a waste product. At the same time they would also use up the carbon dioxide which is produced by the spacemen's respiration and which might accumulate to poisonous levels if not removed.

Figure 58
Neil Armstrong (in the foreground) and Edwin Aldrin rehearsing in April, 1969 for their work on the surface of the Moon later that year. Their backpacks provide them, amongst other things, with oxygen for breathing. *Photograph, NASA.*

How we breathe

5.1 Where the air goes

It is general knowledge that the air taken into our bodies passes to our lungs. It is also mentioned in section 4.3 that this is where gas exchange takes place. It is now time to take a closer look at the route taken by the inhaled air. Information on this can be obtained in a number of ways:

a   A pair of fresh lungs (obtained from a butcher) can be examined and dissected.

b   A liquid plastic can be injected into the trachea (windpipe) attached to a fresh pair of lungs and allowed to flow to fill all the spaces in the lungs. It is then allowed to set. If the lungs are placed in an acid bath to dissolve the soft tissues, what remains is a plastic cast of all the spaces in the lungs connected to the trachea. Many of the finer branches can be 'pruned' to allow the larger ones to be seen. *Figure 59* shows a cast made in this way. (See also colour *plate 3*.)

You can see why the air passages are sometimes called 'the bronchial tree'.

c   An X-ray can be taken of a person's chest after he has inhaled a smoke which is opaque to X-rays. *Figure 60* shows such an X-ray.

From studies such as these, and from others, a picture can be built up of the structure of the lungs and their relationship to other organs in the body. See *figure 61*.

We can now go on to try and answer two major questions:

a   How do we draw air into and out of the lungs?

b   In what part of the lungs does gas exchange take place?

5.2 How is air drawn in and out of the lungs?

If you breathe in deeply, it is obvious that the volume of your chest increases. Two movements bring about this change – those of the ribs and of the diaphragm. If you stand with your hands on your ribs and breathe in deeply, you will feel the movement of these bones.

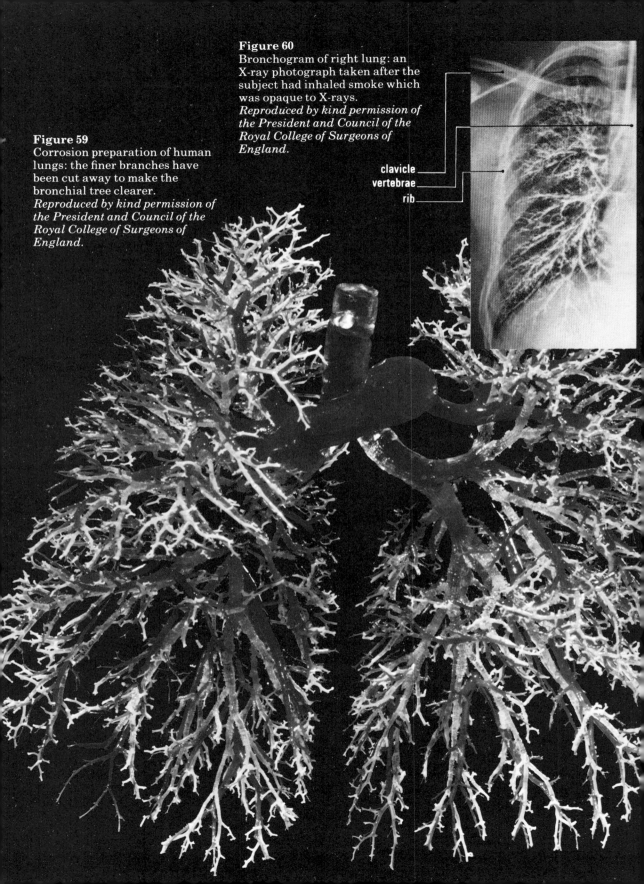

Figure 59
Corrosion preparation of human lungs: the finer branches have been cut away to make the bronchial tree clearer.
Reproduced by kind permission of the President and Council of the Royal College of Surgeons of England.

Figure 60
Bronchogram of right lung: an X-ray photograph taken after the subject had inhaled smoke which was opaque to X-rays.
Reproduced by kind permission of the President and Council of the Royal College of Surgeons of England.

clavicle
vertebrae
rib

Figure 61
Diagram of the human thorax and head showing the route taken by the air passing into the lungs.

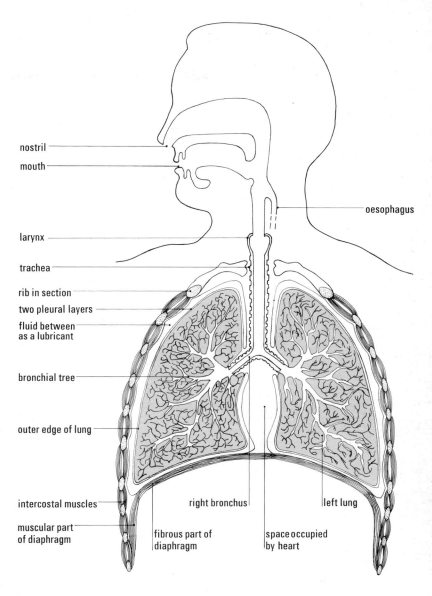

nostril

mouth

oesophagus

larynx

trachea

rib in section

two pleural layers

fluid between as a lubricant

bronchial tree

outer edge of lung

intercostal muscles

muscular part of diaphragm

right bronchus

left lung

fibrous part of diaphragm

space occupied by heart

Q1 In which direction do your ribs move when you are *a* breathing in (inhaling)? *b* breathing out (exhaling)?

These movements are produced by the action of various muscles.

Between each rib and its neighbour run the *intercostal muscles*. Look for them when you next eat a chop – they form the meat running along the edges of the curved rib-bone. These muscles can be seen in the photograph of part of the chest wall (*figure 62a*) and in the drawing (*figure 62b*) which has been traced from the photograph.

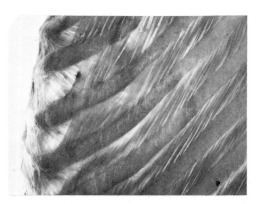

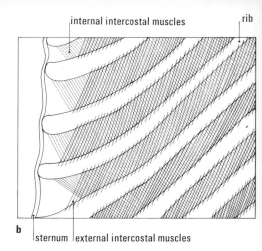

internal intercostal muscles

rib

a

b

sternum | external intercostal muscles

Figure 62
a A photograph of part of the chest wall showing the intercostal muscles running between the ribs.
b A tracing based on *a*.
Reproduced by kind permission of the President and Council of the Royal College of Surgeons of England.

You can get some idea of how these muscles work if you use a simple model like the one shown in *figure 63*. The vertical sides of the model represent the backbone and breastbone (*sternum*). The two nearly horizontal pieces represent two neighbouring ribs. Elastic bands are used to represent the intercostal muscles.

Figure 63
A simple model illustrating the action of the intercostal muscles in moving the ribs.

1 Stretch an elastic band between points P and Q and note what happens.

Q2 Comparing your model with the drawing in *figure 62b* which set of muscles does this band represent?

2 Now stretch the band between R and S and note what happens this time.

Q3 Which set of muscles is represented by the band running from R to S?

Q4 Which muscles appear to cause raising of the ribs and which cause lowering?

Q5 Which muscles do you think cause inhaling and which cause exhaling?

The second movement which increases the volume of the chest is caused by the diaphragm. This is a sheet of muscle and fibre which separates the chest (or thorax) of any mammal from its abdomen. If you look at a dissected mammal you will see that the diaphragm is thin and tough in the centre, but that around its edge, where it is fastened to the body wall, it is thick and muscular. See *figure 64.*

Figure 64
A diagram of the human diaphragm viewed from below.

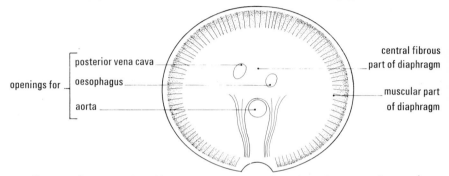

Seen edge on, the diaphragm is not flat, but forms a dome (*figure 61*). The fibrous central portion is the roof of the dome, and the muscular edges form the walls, which in man are almost vertical. You can see the shape of the diaphragm clearly in an X-ray because the liver, which presses up against it from below, stands out on the film (see *figure 65*).

Figure 65
An X-ray photograph of the human chest viewed from the front.
Photograph, Dr R. E. Lawrence, St Mary's Hospital, London W9.

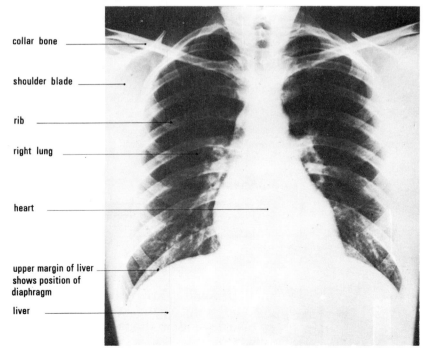

Suppose the muscular part of the diaphragm contracts.

Q6   Which way would you expect the dome to be pulled?

The movements of the diaphragm can be represented in two models which resemble the thorax in certain ways. These are illustrated in *figures 66* and *67*. If you study such models you may find that they do not exactly represent what happens in the actual animal, but at least they help to demonstrate what physical processes occur.

Figure 66
A bell-jar model of the thorax.

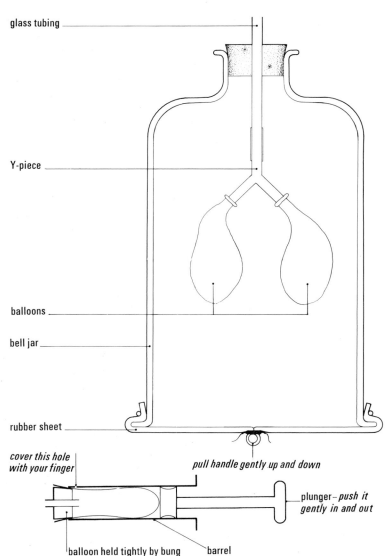

Figure 67
A syringe model of the thorax.

You will probably realize that these models move air in and out by a pumping action. A bicycle pump and other types of pump also depend on the fact that when the volume of a gas is increased, its pressure is decreased, and vice versa.

Living things in action

5.21 Air pressure in the lungs

If the human chest does work in the same way as the two models in section 5.2 we should expect pressure changes within the lungs during breathing.

Q7 What will these pressure changes be as a person *a* inhales and *b* exhales?

It is difficult to measure the air pressure inside the lungs themselves, but it is easier to find out the pressure in the oesophagus (gullet) which is thin-walled and runs through the thorax between the lungs. You can assume that the pressure inside the oesophagus is equal to that in the lungs. Subjects have been asked to swallow a small balloon attached to a device for measuring pressure (a manometer). During normal breathing the pressure in the balloon has been found to fall to some 5 cm of water below atmospheric pressure during inhalation. In a deep inhalation it may fall to as low as 40 cm of water below atmospheric pressure. On the other hand, during breathing out the pressure has been found to rise to about 1 cm of water above atmospheric pressure; this figure may be as high as 30 cm of water during a forced exhalation.

Here, then, is evidence supporting the hypothesis that when the chest starts to expand, the air pressure in the lungs falls slightly below atmospheric pressure, so that air is forced into the lungs. As the chest shrinks the pressure rises and air is forced out again – the person exhales.

Q8 Thinking back can you now say what two actions lead to the expansion of the chest as you inhale?

5.3 The pleural sacs

In section 5.2 we considered movements of the thorax wall which cause changes in air pressure within the lungs. The thorax wall and lungs are separated by two thin, damp membranes. Although *figure 61* shows these two layers with a gap between them, in real life this cavity does not exist – the two pleural layers are only separated by a thin layer of fluid. The pleural layers make up an airtight *pleural sac* around each lung. You will realize the importance of the lungs being situated in airtight cavities if you repeat the 'syringe thorax' experiment (*figure 67*) leaving your finger *off* the hole in the syringe barrel.

Q1 Suppose the pleural sac on one side of the body were punctured. Would breathing be possible in either lung?

Q2 Suppose someone has a punctured pleural sac and chest wall. What sort of first aid do you think might help to restore normal breathing?

5.4 Lung capacity

You will have seen from *figure 61* that much of your thorax is taken up with the two lungs which hold quite a large volume of air.

1 Arrange a graduated bell jar in the way shown in *figure 68*, carefully holding it in a bucketful or sinkful of water. The water should come right up to its neck.

Figure 68
A method of measuring the vital capacity of the lungs.

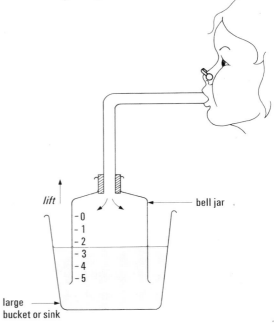

2 One person whose nose is pinched with a nose clip, takes a deep breath and breathes out as much air as he can down the tube. Another person slowly raises the bell jar so that the water levels inside and outside the jar remain the same.

3 When he has breathed out all he can, read off the volume on the graduated scale.
This volume, the maximum that can be breathed out of the lungs after the deepest possible inhalation is called the *vital capacity*.

Q1 What is the point in raising the bell jar as it is filled up?

Q2 Do you think the result you have obtained for the vital capacity represents the *total capacity* of the lungs? Give reasons for your answer.

The amount of air that is moved into and out of the lungs during normal relaxed breathing is clearly much less than the figure for the vital capacity you have obtained. You can measure this *tidal volume* as it is called, in the following way:

1 Hold the bell jar in the water so that the levels inside and out are at some convenient point, for example $2 \ dm^3$.

2 A person with a nose-clip breathes in and out in a relaxed

way while another person notes the upper and lower levels of water in the bell jar. The difference between these levels is the tidal volume.

Q3 What proportion of the vital capacity of the lungs is replaced with each normal breath?

5.5 Artificial respiration

Sometimes, due to accident or illness, a person's breathing may be severely affected and may even cease. It is obviously essential that his breathing should be returned to normal as soon as possible.

People who have had a severe electric shock, or who have been rescued from drowning or gas poisoning, may stop breathing. Their breathing must be started immediately if they are to survive; even a delay of a few seconds may prove fatal, for by then the brain is damaged by lack of oxygen. A rescuer can assist breathing by blowing air directly into the lungs. (See *figure 69*.) As with all first-aid procedures, it is important to know exactly what to do, otherwise you may do more harm than good.

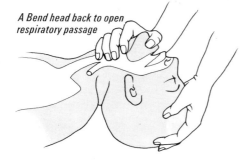

A Bend head back to open respiratory passage

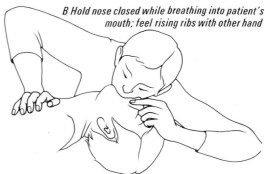

B Hold nose closed while breathing into patient's mouth; feel rising ribs with other hand

Figure 69
Giving artificial respiration by the 'mouth to mouth' method.

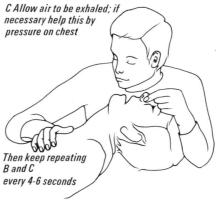

C Allow air to be exhaled; if necessary help this by pressure on chest

Then keep repeating B and C every 4-6 seconds

It would be a good thing if you knew how to apply artificial respiration and the best way to learn is in a first-aid class. Above all, you should not practise this procedure unless you are supervised by a qualified instructor.

5.6 The fine structure of the lungs

Air moving into the lungs passes down the trachea, into one of the bronchi and thence to the bronchial tree. *Figure 59* shows a cast of the bronchial tree made by injecting liquid plastic into the trachea so that it flowed into the finest branches of the lungs. (See also colour *plate 3*.) *Figure 70* shows the same thing in diagrammatic form. Examine particularly the magnified part showing what the tips of the finest branches (the air sacs or *alveoli*) look like.

Figure 70
Details of lung structure at three different stages of magnification.
After Mackean, D. G. (1973)
Introduction to biology,
5th edition, John Murray.

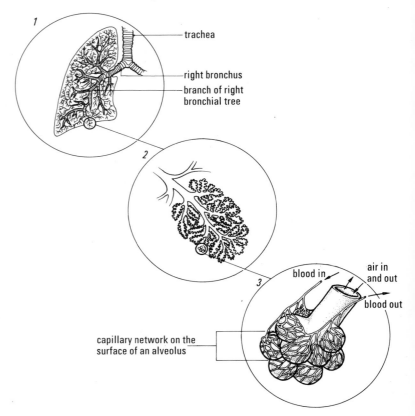

In order to function efficiently in gas exchange, the lungs must possess several important features. They must bring the blood and the air into the closest possible contact and they must do this over as wide an area as possible. Because the lungs are spongy organs, we can see little of their internal structure in sections (thin slices).

Colour *plate 3* shows the intimate relationship of the bronchial tree with the blood system while *figure 70* shows something of the appearance of the fine blood vessels which cover the alveoli. Each alveolus represents a boundary between air and blood; since the two are separated by about a millionth of a metre (1μm), materials can be transferred quickly from one to another. Not only are the alveoli and blood vessels very close to one another; they have a large area of contact. It has been estimated that a pair of human lungs has over 700000000 alveoli, which would give a surface area of about 50 m². You might like to compare this with the surface area of a man's body (about 2 m²) or the floor area of an average classroom (work it out for yourself).

5.7 Gas exchange in the lungs

Your work with gas analysis will have established that animals take in oxygen and produce carbon dioxide. You can learn more about this *gas exchange* by comparing the amounts of oxygen and carbon dioxide in the blood as it enters and leaves the lungs.

Table 8
Volumes of each gas carried by 100 cm³ of blood.

	Blood entering the lungs (cm³)	Blood leaving the lungs (cm³)
nitrogen	0.9	0.9
oxygen	10.6	19.0
carbon dioxide	58.0	50.0

It is also useful to recall the composition of inhaled and exhaled air.

Table 9
Percentage volume composition of various air samples.

	Atmospheric air	Exhaled air
nitrogen	79.01	79.5
oxygen	20.96	16.4
carbon dioxide	0.03	4.1

Q1 Which air sample had the most oxygen in it, and which the least?

Q2 Which blood sample had the most oxygen, and which the least?

Q3 From your answers to questions 1 and 2, what hypothesis best explains what happens to oxygen in the lungs?

Q4 Which air sample had the most carbon dioxide in it, and which the least?

Q5 Which blood sample had the most carbon dioxide in it, and which the least?

Q6 From your answers to questions 4 and 5, what hypothesis best explains what happens to carbon dioxide in the lungs?

The changes you have studied take place in the alveoli. You have seen that these numerous microscopic air sacs present a large surface area to the air we breathe. The gases oxygen and carbon dioxide must be able to pass quickly backwards and forwards between the alveoli and the blood. To do so they must be in solution.

Q7   Can you suggest another feature which the alveoli must possess if they are to operate efficiently in gas exchange?

The surface over which gas exchange takes place is sometimes called the *respiratory surface,* because respiration uses up oxygen and produces carbon dioxide. Similarly our lungs are sometimes referred to as 'respiratory organs' but in fact respiration goes on within every tissue and cell of the body. This process is investigated in Chapter 6. Your studies in the present chapter have revealed something of the structure and function of the lungs, which 'service' respiration by taking in supplies of a raw material (oxygen) and getting rid of a waste product (carbon dioxide).

'Dying for a smoke'

In this country at the present time about two-thirds of all men and one-third of all women smoke regularly. Although tobacco was introduced into Europe in the sixteenth century, smoking only became popular in the United Kingdom during the nineteenth century. You can see in *figure 71* how the introduction of cigarettes at the end of the century affected tobacco consumption.

Background reading

Figures 71 *to 75 are based on The Royal College of Physicians of London (1971)* Smoking and health now, *Pitman Medical Publishing Co.*

Figure 71
Tobacco consumption in the United Kingdom from 1890 to 1968. This figure gives the annual consumption of tobacco in pounds per man or woman aged 16 or over. One pound (lb) = 0.45 kg. 10 lb of tobacco consumed over one year would mean smoking an average of about 15 cigarettes per day.

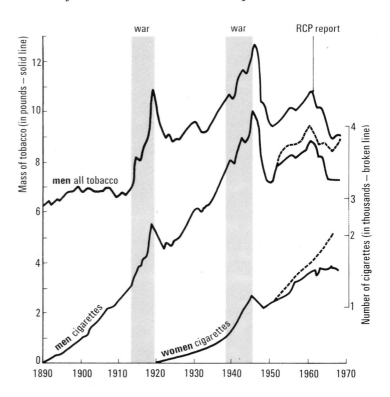

In 1952 two British scientists reported the results of a survey carried out in a large hospital. Patients suffering from cancer of the lung had answered a series of questions – on smoking habits, place of work, type of recreation, method of heating in their homes, and so on. Each lung cancer patient's answers were matched with those from another patient in the same hospital suffering from some other disease; in nearly all cases this patient was of the same sex and roughly the same age, so that the *control* group (patients not suffering from lung cancer) was very similar. When the answers were compared, one difference stood out. *Table 10* shows what it was.

	Non-smokers (%)	Smoking 15 + cigarettes per day (%)
Lung cancer patients	0.5	25.1
Control patients	4.5	13.4

Table 10

Between 1952 and 1964 over 30 similar surveys were done in different countries. All showed the same tendency – lung cancer is correlated with cigarette smoking.

Deaths from lung cancer in the United Kingdom during the 1960s were as follows:

1960	24 800
1965	29 800
1970	34 200

Smoking is associated with several other diseases, chiefly bronchitis and emphysema, coronary heart disease, gastric and duodenal ulcers. Recent studies have shown that unborn babies may be affected by smoking too. Pregnant women who smoke are more likely to give birth to small babies than pregnant women who do not smoke – they are also more likely to have miscarriages. When one considers all possible causes of death, cigarette smokers are at greater risk (shown in *table 11*) than non-smokers.

Table 11
Results of an investigation into the risk of dying, from all causes, during ten-year periods from the age of 35 to the age of 74.
From the Royal College of Physicians of London (1971)
Smoking and health now, *Pitman.*

		Smokers of cigarettes		
Age	Non-smokers	1–14 per day	15–24 per day	25 or more per day
35–44	1 in 75	1 in 47	1 in 50	1 in 22
45–54	1 in 27	1 in 19	1 in 13	1 in 10
55–64	1 in 9	1 in 6	1 in 5	1 in 4
65–74	1 in 3	1 in 2	1 in 2	1 in 2

To take an example, in the group covered, a 50-year-old non-smoker had one chance in 27 that he would die before he was 60; a 50-year-old who smoked 15 to 24 cigarettes a day had one chance in 13.

Some of these facts were first made available to the general public in 1962, when the Royal College of Physicians issued a report, 'Smoking and health'. One of that Report's findings was that cigarette smoking showed a stronger correlation with various diseases than pipe or cigar smoking did. If you look back at *figure 71* you may see that after 1962 there was a slight fall in men's cigarette consumption. There was an even greater change in the smoking habits of *doctors*.

Figure 72
Changes in doctors' smoking habits from 1951 to 1966.

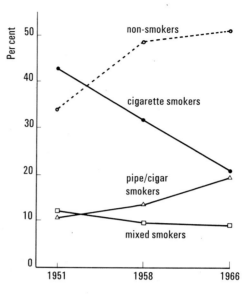

Evidently the very people most aware of the risks in cigarette smoking were doing something about it; the effectiveness of their action is plain when you consider *figure 73*.

Figure 73
Death rates from lung cancer in male doctors and in all men in England and Wales from 1954 to 1965. During this period, many doctors stopped smoking and their death rate from lung cancer declined by 38 per cent while in all men in England and Wales who had not changed their cigarette consumption the rate increased by 7 per cent. This experiment which doctors carried out on themselves is strong evidence of the benefits that would result if many people in all occupations stopped smoking.

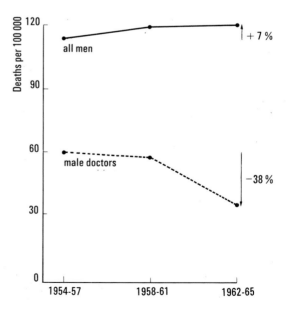

Living things in action

Why do people smoke? Most would say that it soothes the nerves, provides comfort in loneliness or worry and that it stimulates social contacts – to offer someone a cigarette is to make a gesture of friendship. Most people take up smoking in their teens. A survey of adult smokers in 1964 showed that 90 per cent found the habit pleasurable, but 60 per cent said that it cost more than the pleasure was worth. Most people are only vaguely aware of the medical evidence against smoking, but the cost of a packet a day soon adds up. What may have started as a casual pastime at school turns out to be a habit which could cost a 20-a-day man over £100 a year. A smoker may be reluctant to stop because of a mild form of drug dependence on the nicotine he absorbs in his cigarette smoke – certainly many people who give up smoking find that they become restless, crave a cigarette, and frequently put on weight (nicotine tends to lessen the sense of hunger and so a smoker eats less). For some people these *withdrawal symptoms* are too much to put up with and they go back to smoking. For those strong-willed individuals who manage to give it up, there is a big health bonus.

Figure 74
Death rates from lung cancer for cigarette smokers and ex-smokers for various periods, and for non-smokers.

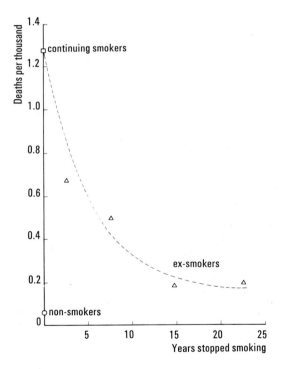

Figure 74 shows that a smoker who gave up two years ago has already halved the risk of developing lung cancer, and the longer he remains a non-smoker the lower this risk will fall. Finally, a further reason why smoking is so popular and giving up so difficult is the large sum spent on advertising. (*Figure 75.*)

Figure 75
Expenditure on advertising
tobacco goods from 1955 to 1968
(adjusted to 1960 costs).

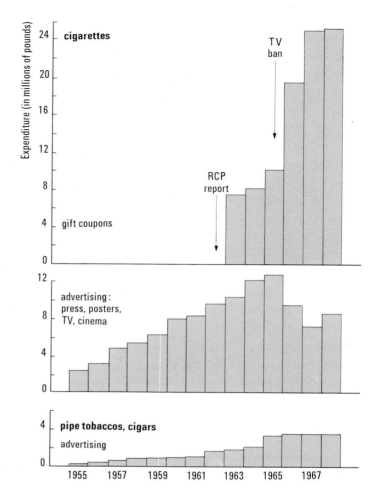

Cigarettes are among the most highly advertised of
commodities, and several steps have been taken to modify
such advertisements. In 1965 there was a ban on
television advertisements for cigarettes, and from April
1971 a printed warning was included on cigarette packets
and on advertisements for them.

One type of promotion that has grown rapidly since its
introduction in 1962 (the year when the report 'Smoking
and health' appeared) has been the gift coupon;
advertisements for coupon brands can point out the
quality of the 'gifts' available, as well as (or instead of) the
quality of the cigarette sold. The gifts are not of course
'free', since their cost is included in the price of the
cigarettes.

'To smoke or not to smoke' is a decision which individuals
make for themselves. But there are a number of solid facts –
financial and medical – which ought to influence that
decision. You may like to think about some of them:

In Great Britain in 1970, 7499 people were killed in road accidents. In the same year 34 200 people died of lung cancer, and over 30 000 from bronchitis and emphysema. In 1968, a total of £1 300 000 was spent on road safety propaganda; in the same year the Health Education Council (a government body) allocated £100 000 to anti-smoking propaganda. Both sums are dwarfed by that year's expenditure on tobacco advertising; about £52 000 000.

Cigarettes cost a lot not only because they are heavily advertised but also because they carry high rates of tax and import duty. In fact, tobacco provides a sizeable part of the government's revenue – about £1 000 000 000 every year. So, although the country spends a great deal on tobacco (in 1969 more than it spent on meat and bacon, or on heating and lighting the home) much of the money finds its way into the Exchequer.

Q1　Suppose that the Chancellor of the Exchequer could raise a thousand million pounds a year by some other tax, would it be a good idea for the government to pass a law against cigarette smoking?

Q2　How far can anyone say, 'no one has *proved* that cigarettes cause certain diseases. There is simply an association'?

Organisms at work

6.1 Energy from food

You will have seen in Chapter 4 that all organisms produce
carbon dioxide. They do this as a result of using up food
within their cells. It is difficult to provide experimental
evidence for this in school but a suitable experiment is
illustrated opposite. Food contains various substances.
When these substances are broken down, inside living cells,
energy is made available and carbon dioxide is released. It
is almost as if the food is the 'fuel' which 'powers' the
organism. The fact that food can release energy can be
demonstrated in quite a dramatic way. The apparatus is
shown in *figure 76*.

Figure 76
Apparatus for burning dry
custard powder.

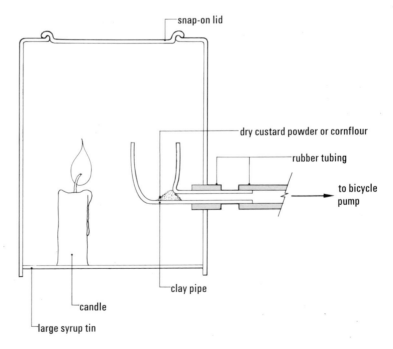

The bicycle pump is used to produce a cloud of custard
powder which is then ignited by the candle.

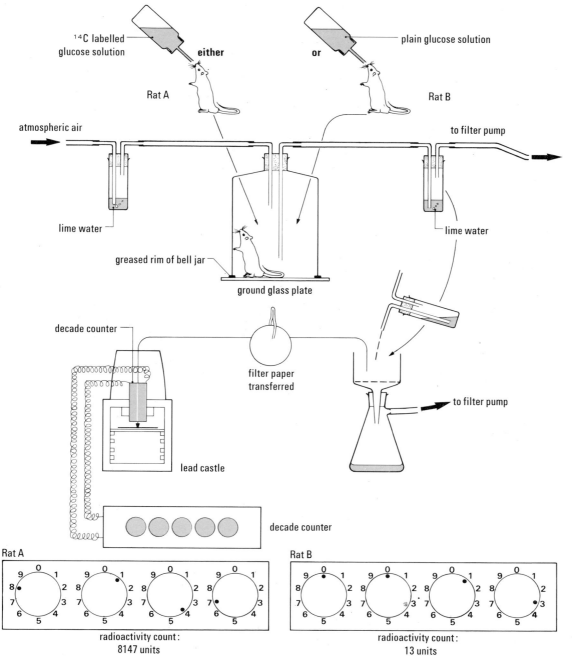

Figure 77
Feeding rats with ^{14}C-labelled glucose, and testing their exhaled carbon dioxide for radioactivity.

You can now go on to investigate the energy released from a burning peanut.

1 Pour exactly 20 cm^3 of water into a boiling-tube. Clamp the tube to a retort stand.
2 Measure and write down the temperature of the water.
3 Weigh a peanut and then, carefully, but securely, impale the nut on a mounted needle.

4 Hold the nut in a Bunsen flame until it starts to burn, then immediately place it under the boiling-tube so that the heat from the burning nut warms the water. (See *figure 78*.)

5 Allow the nut to burn completely beneath the boiling-tube. If it goes out quickly (for instance because of a draught) light it again.

6 When the nut has burned away completely and gone out, take the temperature of the water.

Energy is measured in *joules* (J). You can calculate the actual amount of energy involved in heating the water if you know

a the mass of water involved (in this case 20 g), *and*

b the rise in temperature through which it was heated. The calculation is quite simple:

$$\frac{\text{Energy released}}{\text{from nut (in J)}} = \text{mass of water} \times \text{temperature rise} \times 4.2$$

This may come to rather a large figure and so for convenience it is usual to express the energy in kilojoules (kJ). One kilojoule equals 1000 joules.

Figure 78
Apparatus for investigating the energy released from a burning peanut.

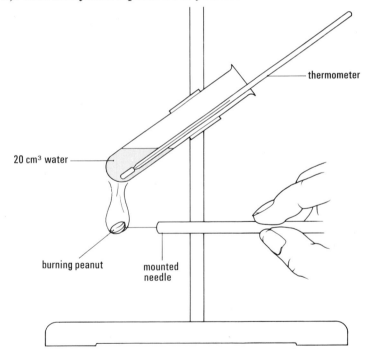

thermometer

20 cm³ water

burning peanut

mounted needle

When slimmers talk of the energy content of food they usually refer to calories rather than joules. In fact this 'calorie' is really a kilocalorie and is equal to 4.2 kilojoules.

If you know the mass of the nut you used, you can extend the calculation to working out the energy released from *one gram* of peanut. You will then be able to compare your results with those of others in the class.

Living things in action

Q1 How variable were the results of the class for the energy value of 1 gram of peanut?

Now compare the results you have obtained with the 'official' value in the table in the Appendix (page 245). Remember the value in the table is given per 100 g of peanut. You will probably see that your figure is rather different from that in the table. Look back at the procedure you used and bear in mind that to obtain an accurate value

a   *all* the nut must be burned
b   *all* the energy given off by the burning peanut must be used to heat up the water in the boiling-tube.

Q2 List as many ways as you can in which your experimental procedure may have led to inaccurate results.

Q3 Can you design a set of apparatus which would avoid as many as possible of the sources of error you have listed in your answer to question 2?

6.2 The release of energy by living organisms

In living animals and plants, energy is not released by burning. The process is similar to burning in some ways, in that food substances are made to combine with oxygen, breaking down to carbon dioxide and water. However, the process takes place more slowly and in a more controlled way than it does in burning. This reaction, which continues in *all* living cells, also releases energy and is called *respiration*.

$$\text{Food} + \text{oxygen} \nearrow \text{energy released}$$
$$\searrow \text{carbon dioxide} + \text{water}$$

This energy, which living organisms obtain from their food, is used in various ways. In many cases it is converted into another form of energy, such as one of those mentioned below.

6.21 Light energy

The 'glow worm' is really the wingless female of a species of beetle. Her light probably attracts males. Luminescent animals can also be found in the depths of the sea where some, like the angler fish, may use luminous lures to attract their prey. Luminescent organisms are relatively common in tropical countries where the night may be lit by the flashes of flying insects, such as the fire-fly. Some plants, especially certain species of bacteria and fungi, also produce light as they break down organic material such as logs or the bodies of animals. (Colour *plates 4a* and *b*.)

6.22 Electrical energy

The electric eel generates a sufficiently high voltage to stun its prey. In fact electrical activity can be detected in most living organisms, though at a much lower voltage than in the electric eel. The human brain shows such activity, though this can only be detected with very sensitive electrodes. See *figure 79*.

Figure 79
An electroencephalograph machine in operation.
Photograph by courtesy of Dr W. A. Kennedy, Brook General Hospital, London SE18.

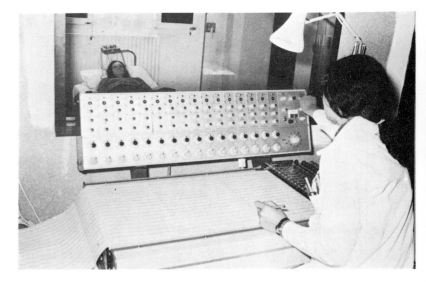

6.23 Sound energy

Many animals can make sounds of one kind or another. This is usually done by producing vibrations at a particular frequency. In the human voice it is the vocal cords which are made to vibrate as air passes over them. In a grasshopper the vibrations are produced by moving a leg against the side of the body.

6.24 Heat energy

Burning releases nearly all its energy as heat. Although respiration oxidizes food in a much slower way than burning, one can detect heat output in many cases.

1 Use a clinical thermometer to take your own temperature and compare it with that of your surroundings.
2 Compare your body temperature with that of other members of your class.

Q1 How much variation in temperature is there among the members of your class?

Q2 From your observations, what do you think is meant by the term 'normal body temperature'?

It is well known that mammals, such as ourselves, can maintain their temperature at a fairly constant level. (See Chapter 3.) The question now arises as to whether other living organisms – plants and insects, for instance – also produce heat.

Q3 Can you design an experiment to find out whether small plants (such as germinating seeds) or insects (such as maggots) produce heat? Bear in mind that the amount of heat produced, if any is produced at all, may be very small.

6.25 Chemical energy

Within our bodies we use energy, made available to us through respiration, to produce complex chemical substances. Thus, the energy contained in the food we eat may be used later in the production of new muscles, hair, and bone. This energy 'locked up' in complex substances is sometimes called chemical energy. Chemical energy may be difficult to visualize, but its use in the body of any living organism never stops.

6.26 Movement energy

When an object moves from one place to another, energy is used. This movement energy is sometimes called mechanical or *kinetic* energy. Whatever form the movement takes and whether it involves a living or a non-living thing, the energy enabling the movement must have come from somewhere else.

Figure 80
One way of getting uphill.
Photograph, E. D. Lacey.

Compare the two photographs in *figures 80* and *81*. In both cases kinetic energy is being used.

Figure 81
Another way of getting uphill.
Photograph, E. D. Lacey.

Q4   What provides the energy for the movement in each of these cases?

Q5   Make a list of the similarities between the two ways in which chemical energy is being converted to kinetic energy.

6.3 Measuring human energy output

You have spent a lot of time in this chapter looking at energy in its various forms. Energy can be described as 'the capacity for doing work'. *Figure 80* shows an instance of a person doing work. He has to apply a considerable force to raise his body (and the bicycle) up the hill – and, of course, to overcome wind resistance and friction too. If we could measure the work which he is doing, then we would have some idea of the energy output of which he is capable.

An ergometer is a piece of apparatus designed to do just this. Ergometers are of various designs – one based on a bicycle is shown in *figure 82*. Other ergometers can be used to find the energy output of the arm, rather than the legs.

Figure 82
Using a bicycle ergometer.
Photograph, Michael Plomer.

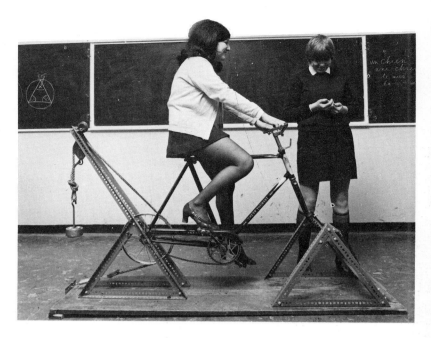

Alternatively, a person's rate of energy output can be assessed by measuring the rate at which he can run upstairs or lift weights from the floor onto a table.

Use one or more methods to find out the rate of energy output by human subjects.

Compare your results with those of other members of the class; then try to answer the following questions:

Q1 How much variation was there in the rates of energy output of various people, measured by the same technique? Can you suggest a reason for this?

Q2 How much variation was there in the rate of energy output by the same person, measured by different techniques? Can you explain this?

6.4 Energy conversion in living organisms

By now you will realize that living organisms expend energy in a variety of ways. You have also seen that when food is oxidized to carbon dioxide and water, energy is released. If energy is released when the complex substances in food are broken down to simpler ones, then it follows that energy must have been needed to form these substances in the first place. Living organisms are thus able to convert energy from one form into another. The source of an animal's energy is therefore the chemical energy in the food which it eats. Fuels such as coal, oil, natural gas, and wood, also release large amounts of energy when they are burned. Energy was originally involved in promoting the growth of the trees and other plants which formed these fuels, sometimes millions of years ago. Therefore the energy of the 'fossil fuels' – coal, oil, and gas – is directly derived from complex substances made by living organisms. The way in which this energy is 'locked up' in complex substances is something to which you will return in Chapters 10 and 11.

6.41 Energy conversions within the living cell

The cells of living organisms continuously use energy derived from food. In whatever form the food is eaten, much of it eventually reaches the cells as a sugar called *glucose* (see page 103). The chemical energy released during the breakdown of glucose is then made available to the cells by a complex mechanism. This involves a substance called *adenosine triphosphate*, usually abbreviated to ATP. ATP acts as a vital intermediate in the transfer of energy, derived from the glucose through respiration, to the various energy-requiring processes of the living cell.

6.5 Muscles in action

When animals move, they use their muscles. These are the organs which carry out the conversion of chemical energy to movement (kinetic) energy. We can study the effect of certain substances on muscle fibres.

1   Take a clean, dry microscope slide and a fine strand of muscle fibres (a narrow strip of lean meat).
2   Put the slide on your bench and carefully arrange the muscle fibres on it so that they are perfectly straight. Then measure their total length in mm. (See *figure 83*.)
3   Add a few drops of distilled water to the fibres. Allow about a minute for the water to soak in, then measure the new length of the muscle fibres as accurately as you can; record the length in mm.
4   Tilt the slide to drain off the excess water, and use a piece of filter paper to soak up the liquid around the muscle.
5   Now add a few drops of glucose solution. Leave for about a minute and measure the muscle's length.
6   Drain away the excess solution as before. Now add a few drops of ATP solution; leave for about a minute and then measure the length of the muscle fibres once more.

Think about your results. It may be useful to see what results other members of your class have obtained.

Q1   What is the effect of each of the liquids on the muscle fibres?

Figure 83
Testing the effects of different liquids on the length of muscle fibres.

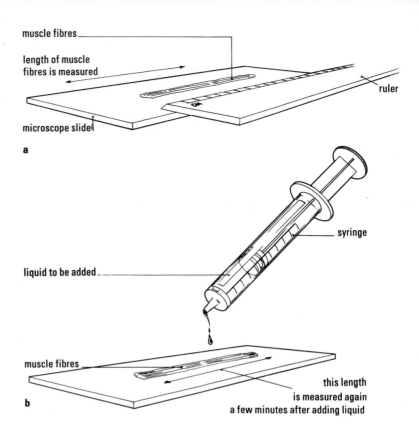

muscle fibres

length of muscle fibres is measured

ruler

microscope slide

a

syringe

liquid to be added

muscle fibres

this length is measured again a few minutes after adding liquid

b

Q2   The experiment consists of adding three liquids in turn to a piece of muscle. Can you think of one way in which this procedure could be criticized as unreliable? How would you modify the procedure to meet this criticism?

Q3   Why is it important to find the effect of water on the muscle?

Q4   Why do you think a long, narrow strip of muscle, consisting of few fibres, was used for the experiment?

6.6 Respiring without oxygen

The energy for a process like muscle contraction comes from food. You have seen that the food, such as glucose, is made to combine with oxygen.

$$glucose + oxygen \underset{\longrightarrow}{\overset{\longrightarrow}{}} \begin{array}{l} energy\ released \\ carbon\ dioxide + water \end{array}$$

This process is called *aerobic respiration* as the oxygen normally comes from the air. As you sit reading this book, the cells in your body are respiring aerobically. Aerobic respiration occurs in plant cells as well as those of animals. However, sometimes energy can be released from glucose in the complete absence of oxygen. This type of breakdown –

called *anaerobic respiration* – takes place in muscles which are very active, for instance during running.

glucose $\overset{\displaystyle\longrightarrow}{\displaystyle\longrightarrow}$ some energy released / lactic acid

The lactic acid produced during anaerobic respiration accumulates in the muscles. After the activity has stopped, the lactic acid is removed by oxidation. You can investigate the effects of this lactic acid on muscles with a simple experiment.

1 Raise one arm above the head and lower the other.
2 Clench and unclench each fist. Work both hands at the same rate, preferably about twice every second.
3 Notice the sensation in each arm as the exercise goes on.
4 When one arm tires, rest both arms on the bench in front of you and notice what you feel in each.

Q1 Which arm tired first?

Q2 What were the sensations of 'tiredness'?

Q3 What were the feelings in the tired arm when it was rested in a horizontal position?

Q4 Can you explain why one arm tired so much more quickly than the other?

This experiment, and the answers you have given may enable you to answer the next two questions.

Q5 Why does a good sprinter not breathe at all during a 100 m race?

Q6 Why, after the finish of an 800 m race, do runners pant very hard for several minutes?

6.61 Anaerobic respiration in yeast Yeast is a micro-organism which, like your muscles, can respire in two ways. You can investigate some aspects of this in a simple experiment. The indicator, diazine green, changes colour according to the amount of oxygen present:

Oxygen present   *Oxygen absent*
Indicator oxidized ↔ Indicator reduced
to *blue* colour   to *pink* colour

1 Place 20 cm³ of glucose solution, with some yeast added to it, in a test-tube. Add two drops of diazine green solution. The colour produced will tell you whether oxygen is in the mixture.
2 Pour enough liquid paraffin over the mixture to form a layer right over the surface. This layer will prevent more

oxygen from the air from getting into the mixture.

3 As soon as the colour of the mixture indicates that there is no oxygen present, fix up a delivery tube leading to a small amount of bicarbonate/indicator in another test-tube. (See *figure 84*.)

Figure 84
Investigating whether yeast can respire anaerobically.

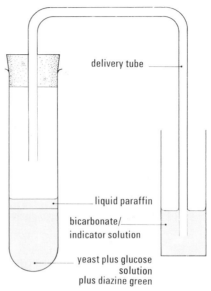

delivery tube

liquid paraffin

bicarbonate/
indicator solution

yeast plus glucose
solution
plus diazine green

4 Design and set up a suitable control. Remember that in this experiment you are trying to find out whether the yeast is respiring in the absence of oxygen.

5 Leave the apparatus for 10–20 minutes and then record your observations.

Q7 What can you conclude from the final colours of *a* the bicarbonate/indicator solution, and *b* the diazine green solution?

Q8 What can you conclude from your two answers to question 7?

6.62 Fermentation

If yeast is left in a sugary liquid for some days, a characteristic smell is produced. This is the result of ethanol accumulating in the mixture. *Fermentation* is the name given to this process in which yeast turns sugar into ethanol:

Sugar → ethanol + carbon dioxide

Q9 Suggest a way of discovering whether any energy is released during this process.

1 Prepare two balls of cottonwool, each about 1 cm across.
2 Place one on an asbestos mat. Light it with a match and time how long it burns.

3 Place the second ball on the mat. Add two drops of ethanol to it. Light this ball and time how long it burns.

Q10 Apart from the duration of burning, was there any difference in the burning of the two balls of cottonwool?

Q11 What does this experiment tell you about the energy conversions during the anaerobic breakdown of sugar to ethanol?

Q12 Can you suggest a reason why most cells do not respire anaerobically at all times?

Respiration and athletics

When an athlete starts to run, the stores of available energy in his muscle cells are used up in making his muscles contract. In order to continue using his muscles the athlete must release more energy as fast as he requires it. At first he does this aerobically. How fast he can do it depends on how fast he can supply the muscles with oxygen. A fit athlete can absorb about 4 dm^3 of oxygen per minute. This will release about 80 kJ of energy. However, not all of this is converted into kinetic energy to keep the athlete moving. About 80 per cent of it is released in the form of heat or is used in overcoming internal friction. As a result the energy released by the 4 dm^3 of oxygen per minute will enable an athlete to run at about 20 km per hour.

If an athlete could keep up this speed for over two hours he would have a good chance of a medal in an Olympic marathon race. To win a sprint event, however, he would have to run nearly twice as fast. He obtains the energy for this extra speed from anaerobic respiration. This more than doubles the energy available to the athlete, but it cannot continue at this rate for more than a short time. After this, the accumulation of lactic acid reaches an intolerable level.

After an intense effort, requiring anaerobic respiration, the athlete continues to breathe heavily to obtain oxygen to convert the lactic acid to harmless products. This extra amount of oxygen required is called the *oxygen debt*. An athlete can incur a maximum oxygen debt of around 17 dm^3. It will take him about 45 minutes before this debt is fully repaid and his breathing returns to normal. *Table 12* shows how much athletes in various events use the powerful energy boost provided by anaerobic respiration. The figures are approximate. Oxygen requirements will vary with the size, shape, and physiology of the individual.

Figure 85 *(page 97)*
Explosive events.
a Modupe Oshikoya in the 100 m hurdles in the 1974 Commonwealth Games, New Zealand.
b Bob Beamon (U.S.A.) winning a Gold Medal for a world record long jump (29 feet 2½ inches = 8.9 metres) at the Olympic Games in Mexico in 1968.
Photographs: **a** *Mark Shearman;* **b** *E. D. Lacey.*

Living things in action

Event	Speed of top athlete (kph)	Total energy expended (kJ)	Oxygen needed (dm³)	Oxygen breathed in (dm³)	Oxygen debt (dm³)	Percentage of energy from: aerobic respiration	anaerobic respiration
100 m race	37	200	10	0–0.5	9.5–10	0–5	95–100
800 m race	27	520	26	9	17	35	65
1500 m race	25	720	36	19	17	55	45
10 000 m race	21.5	3000	150	133	17	90	10
Marathon (42 186 m)	20	14 000	700	685	15	98	2

Table 12
A breakdown of the approximate oxygen requirements of winning athletes in different kinds of events.

Training

You have seen that no oxygen is required to release the energy required for a short sprint. The same is true of jumping and throwing events. These are sometimes referred to as 'explosive events' (see *figure 85*). The

athlete's performance in these events depends, not on his
capacity for respiration, but on the strength of his muscles
in relation to his body mass. It also depends, of course, on
the skill with which he uses his muscles. All the forms of
strength training for explosive events depend on the
principle that muscles respond to hard work by becoming
stronger.

In 800 m and 1500 m races ('middle distance' events)
athletes derive about half the energy they require
aerobically and the other half anaerobically. (See *table 12.*)
If they run too fast during the race, their oxygen debt
reaches an intolerable level before the end. A middle
distance runner may enter the finishing straight with a
good lead, only to find that lactic acid slows his muscles
till he is passed by other runners who have judged their
effort better. (See *figure 86.*) The good middle distance
runner judges his speed so that the combined energy
supplied by aerobic and anaerobic respiration is enough
to get him to the finish just as he reaches his maximum
oxygen debt.

Figure 87
L. Viren of Finland has the ideal physique for a long distance runner since he is lightly built and has an exceptional capacity for oxygen intake. This photograph shows him in the 10 000 m race in the Olympic Games at Munich in 1972, which he won.
Photograph, E. D. Lacey.

Middle distance runners must train for both speed and stamina. For speed they train like sprinters, while for stamina they go for long runs to improve the ability of the heart and lungs to supply oxygen to the muscles. They also employ interval training: repeated fast runs of a half to two minutes' duration with a brief rest between each run. Each run increases the lactic acid level in the body. The muscles seem to adapt themselves to working in these conditions and so the athlete tolerates a greater oxygen debt.

In long distance races, anaerobic respiration can supply only a very small proportion of the total energy requirement. Most of the race is run at a fairly even pace which corresponds to the fastest rate the athletes can maintain, using the energy derived from aerobic respiration alone. Hence a top-class 10 000 m race is commonly run at a pace of 65–70 seconds per lap except at the end. (The last lap may take only 50–55 seconds.) A good long distance runner (see *figure 87*) must therefore maintain a high rate of aerobic respiration by increasing the depth and frequency of his breathing. However, this will be of little use unless the increased oxygen is transported to the respiring muscles where it is needed. For this reason modern long distance runners concentrate on training methods which make heavy demands on the heart and circulatory system. They have found that just as other muscles get bigger and stronger through doing a lot of heavy work, so the heart gets large and strong if frequently called upon to work hard.

Half a century ago athletes were only 'in training' for a few months each year. Their training involved little more than brisk walks and short runs. Since then athletes have shown that the body can respond to abnormal demands made upon it by changing in such a way as to make it better able to meet these demands. Their training continues throughout the year and its nature has changed enormously. Many athletes now run weekly distances of 200–300 km – the distance from London to Birmingham and back. These athletes have been rewarded with new world records. In 1922 the world record for the 10 000 m was held by 'the flying Finn', Paavo Nurmi, with a time of 30 minutes 40.2 seconds. Fifty years later the world record stood at 27 minutes 38.4 seconds and times of less than 29 minutes are commonplace. Physiologists have found that these athletes, through their training, have not only strengthened their hearts, but also increased the number of red cells which carry oxygen in the blood. They have also brought about various other changes which have increased the efficiency of their breathing and circulation.

Athletics and altitude

When it was announced that the 1968 Olympic Games were to be held at an altitude of 2000 m in Mexico City, scientists were quick to point out that the reduced atmospheric pressure (see *table 7* on page 63) would have a considerable effect. They suggested that performances in the 'explosive events' would improve because of decreased air resistance. But they also said that the shortage of oxygen would lower performances in long distance events. A world-famous physiologist even said, 'there will be those who will die'.

A team of British athletes and scientists went to Mexico City for a month in 1967 to try and find what the effects of altitude might be. Before they left England they underwent laboratory tests and running trials. These were repeated during their stay in Mexico. *Figure 88* shows the results.

Figure 88
Results of a series of races of over 5 km held in 1967 to compare the performance of British athletes in England and in Mexico.

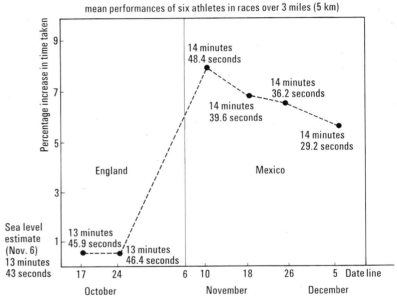

mean performances of six athletes in races over 3 miles (5 km)

The sea level races were run in autumn on tracks in bad condition. It was estimated that the athletes would have run about 3 seconds faster in normal condition

Four days after arriving in Mexico their times were 8 per cent slower than they had been in England. Their times improved throughout their stay, but they were still 5 per cent slower at the end of the month. The improvement was thought to be due to *acclimatization*. Within a few days of arriving at a high altitude, people adapt to the oxygen shortage by breathing faster and deeper. After a few weeks the red cells in the blood become more numerous and there is an increase in the total volume of blood in circulation. As a result of these experiments it was recommended that long distance runners in the British team should acclimatize in Mexico for as long as possible in advance of the Olympic Games.

Living things in action

During the Mexico Olympics competitors in 'explosive events' performed uniformly better than usual. Many world records were broken by wide margins. In the middle distance events, athletes benefited from lower air resistance, but suffered from shortage of oxygen. These two factors resulted in their times being about normal. In the long distance races, however, times were markedly slow; no athletes died, but a number collapsed. (See *figure 89.*)

Figure 89
Martin Winbolt-Lewis (Great Britain) receiving oxygen after a 4 × 400 m relay in the Olympic Games in Mexico in 1968. *Photograph, Mark Shearman.*

Perhaps the most significant fact was that many of the medal winners in the long distance races had lived and trained at a high altitude for most of their lives. Altitude-trained athletes filled the first four places in the 10 000 metres.

Results of 10 000 metres, Mexico City Olympic Games 1968
1 Temu (Kenya)
2 Wolde (Ethiopia)
3 Gammoudi (Tunisia)
4 Martinez (Mexico)

An examination of people living permanently at high altitudes shows that they frequently possess those adaptations to their oxygen transport system which long distance runners acquire as a result of prolonged hard training. Since the Mexico Olympics highlighted this situation, many top class athletes from lowland countries have prepared for major races with long training periods at altitude. Thus the choice of an unusual venue for the Olympic Games stimulated the application of biological principles to training. In the future, science will undoubtedly make an increasing contribution to the raising of standards in sporting achievements at the highest levels.

Food

7.1 Why we need food

No animal can live without food. To understand why this is so, you will need to know what kinds of food there are, and the use to which each is put in the body.

The work on respiration in Chapter 6 suggests that some food is used as *'fuel'*. No part of the body can survive without a supply of food to provide the energy needed for its activity. In every living cell, food is oxidized to release this energy, and most of your food is used in this way.

Food is also needed for *growth*, and any shortage may slow down or even prevent growth altogether. (See *figure 90*.)

Figure 90
Three babies: the one on the left does not get enough food of any kind; the one in the middle lacks protein; the one on the right is well nourished.

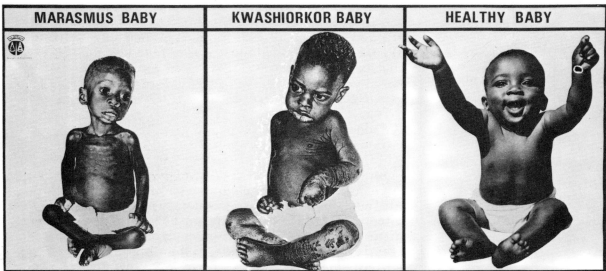

| MARASMUS BABY | KWASHIORKOR BABY | HEALTHY BABY |

From a poster published by the National Food and Nutrition Commission of Zambia, reprinted in Britain by the Voluntary Committee on Overseas Aid and Development.

Adults may have stopped growing, but they still need food for repair and replacement.

Some bodily processes require particular substances. Calcium salts are required, for instance, for the building of bones and teeth and for the formation of a blood clot when the skin is cut.

Animals must therefore eat for a variety of reasons and illness can occur if essential items are missing from the diet.

7.2 Different kinds of food

Food and eating habits vary greatly in different parts of the world. What people eat is more often determined by what they can afford than by what they would like. No matter what the local food may be, the diet must contain certain essential food substances. These are carbohydrates, fats, and proteins. To these we must add certain other nutrients – vitamins and mineral salts – which are required for specific purposes. We must also include water because large volumes of water are lost from the body every day and have to be replaced. (See Chapter 12.)

Only a few foods, such as sugar, are single chemical substances. Most of what we eat consists of mixtures of simple chemicals, such as water and salts, with more complex substances such as carbohydrates, fats, and proteins.

A diet which meets all the essential requirements of a human (or other animal) is usually called a *balanced diet*; it might be more sensible to call it a *complete diet*. To understand the part played by different nutrients you must study each in some detail.

7.21 Carbohydrates

In many societies, such as our own, some children and adults overeat; they put on fat and become overweight. Most people know that to weigh less you must eat less carbohydrate. This group of food substances includes the sugars and starch present in so many of the things you eat.

Chapter 6 discusses the way in which living cells release energy from a sugar called *glucose*. Its chemical formula, $C_6H_{12}O_6$, illustrates the point that all carbohydrates contain the elements carbon, hydrogen, and oxygen. They do not, as the name seems to suggest, contain water, but the hydrogen and oxygen, as in water, are in the proportion of two to one.

The most common carbohydrate in our food is *starch*. You can compare some of the properties of glucose and starch in a simple way.

1 Taste a small quantity of dry, powdered glucose. Compare the taste of dry starch.

Q1 What do you notice about the taste of these two carbohydrates?

2   Place about 10 cm³ of water in a test-tube. Add a *little* powdered glucose. Mix the glucose with the water to find out if it is soluble.

3   Repeat *2* using starch instead of glucose.

q2   What can you conclude about the solubility of glucose and starch?

The chemical difference between sugars and starch can be shown by two simple tests. Record your results in the form of a table.

Test for starch

1   Put a little dry starch on a white tile or watch-glass.
2   Add to it a few drops of dilute iodine solution. Note what happens.
3   Repeat the test with a little glucose instead of starch.

q3   In what way did the reaction of starch with the dilute iodine solution differ from the reaction of the glucose?

Test for reducing sugars

1   Put a small pinch of glucose in a test-tube and add about 1 cm³ of water. Shake (and warm if necessary) to dissolve the glucose.
2   Add about 1 cm³ of Benedict's solution and bring carefully to the boil by placing the test-tube in a beaker of boiling water.
3   Continue boiling for a few seconds and note any colour changes.
4   Repeat the test using a pinch of starch. Note any colour change.

q4   In what way did the glucose and starch show different reactions when heated with Benedict's solution?

7.22 Carbohydrate molecules

Glucose is the simplest abundant sugar and its 24 atoms are joined to one another in a quite definite pattern (see *figure 91*). Starch molecules are large because they are made up of long chains of similar units, each of which is like a glucose molecule (see *figure 92*). One chain may have several hundred such units. Cellulose, which is the chief material in plant cell walls, is another carbohydrate with big molecules made of even longer chains of glucose units than starch.

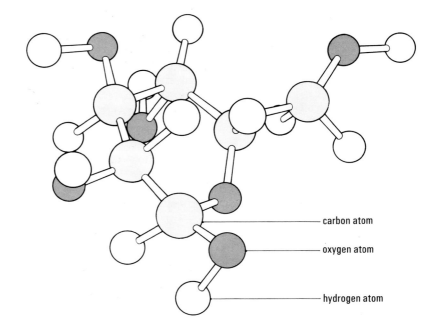

Figure 91
A ball and rod model of a glucose molecule.

carbon atom

oxygen atom

hydrogen atom

Figure 92
A diagram of part of a starch molecule, consisting of a large number of glucose units joined together.

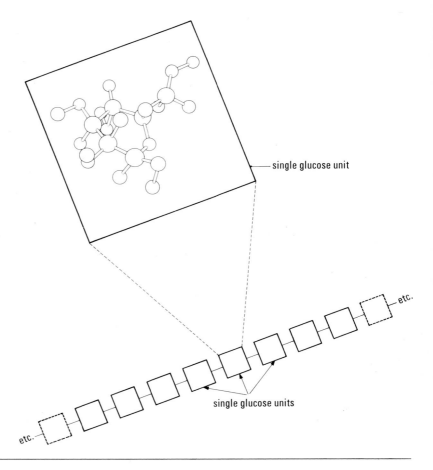

single glucose unit

single glucose units

etc.

etc.

7.23 Fats

An experiment in Chapter 6 measures the amount of heat produced by burning a peanut, which contains a lot of fat. Peanuts burn with a rather smoky flame showing that they contain a lot of carbon. Fats contain the same three elements – carbon, hydrogen, and oxygen – as carbohydrates. However, in fats the proportion of oxygen is low. Tristearin for instance (the fat found in beef and mutton) has the formula $C_{51}H_{98}O_6$.

Fats often have a slippery, greasy feel. They make a translucent stain on paper when they come into contact with it. This feature (seen when chips are wrapped in paper) is sometimes used in testing foodstuffs for fat. You can try this for yourself, by rubbing a piece of fat on glazed paper.

Q5   Can you devise a suitable control for this test?

Fats do not readily mix with water, as you will know if you have rinsed a greasy plate under a tap. But if the fat is broken down into very tiny drops, these will spread throughout the water to form a mixture called an *emulsion*. Milk is a good example of this. Fats will dissolve in a number of liquids other than water, *e.g.* ether, petrol, dry cleaning liquids, and ethanol. Of these, only ethanol mixes readily with water; this property is used in testing foodstuffs for fats.

Test for fat
1   Put about 2 cm³ of ethanol in a test-tube and add *one drop* of olive oil. Shake to dissolve.
2   Almost fill a second test-tube with water.
3   Carefully add the fat in ethanol solution drop by drop to the water in the second test-tube. (See *figure 93*.) Note what happens.

Q6   Can you devise a control for this test?

Q7   Explain what happened to the fat as the solution was poured into the water.

Figure 93
The emulsion test for fats.

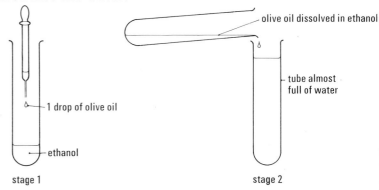

olive oil dissolved in ethanol

tube almost full of water

1 drop of olive oil

ethanol

stage 1   stage 2

7.24 Proteins

Water is by far the commonest substance in animal and plant cells, but the main *solid* component is often protein. Cell growth and all the activities going on in cells depend on protein. Proteins are chemically more complex than carbohydrates or fats; they, too, contain the elements carbon, hydrogen, and oxygen with, in addition, nitrogen and smaller amounts of sulphur and phosphorus.

Test for protein

1 Place about 1 cm^3 of a protein solution or suspension in a test-tube.

2 Add about 2 cm^3 of Millon's reagent. *This solution must be handled with care as it is poisonous.*

3 Bring the mixture to the boil by placing the tube in a beaker of boiling water and note any changes you see.

4 Repeat the test, using first glucose and then starch solution or suspension.

Q8 What colour change was produced by the protein and Millon's reagent which was not produced by the glucose or starch?

7.25 Protein molecules

Figure 94
A very small part of a protein molecule. Each shape represents a different amino acid. Each is also numbered to distinguish it from the others.

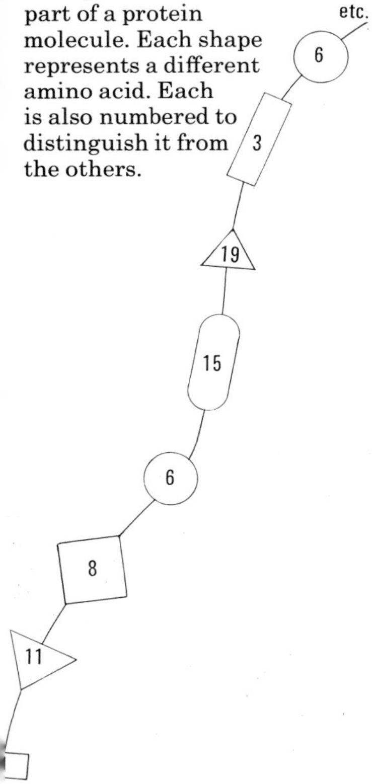

In one animal or plant, *e.g.* in the human body, there are thousands of different proteins, and no two types of organism have precisely the same proteins. In spite of this great variety, all proteins are made of the same units, called *amino acids*. There are altogether about 20 different amino acids, and these are linked in chains to form protein molecules.

The chains are sometimes folded up or twisted round each other, and are often cross-linked by bonds between adjacent chains. Protein molecules contain any number from 50 to several thousand amino acid units – any particular amino acid being repeated many times along the chain. They are therefore very large molecules and the number of different kinds of protein is enormous.

Humans can make half of the 20 different amino acids they need, but there are ten which they cannot make. These are called *essential amino acids* because they must be obtained from food. Food which can supply all the essential amino acids is sometimes called 'first class protein', *e.g.* animal protein such as that in meat, fish, eggs, and milk. Plant protein often lacks certain essential amino acids and may be described as 'second class protein'. The effect of depriving a mammal of just one of the essential amino acids is dramatically shown in *figure 95*.

Figure 95
a The effect of leaving one essential amino acid out of a rat's food.
b The same rat 25 days after the amino acid had again been included in the diet.
From Rose, W. C., and Eppstein, A. (1939) J. biol. chem. **127**, *677.*

a

b

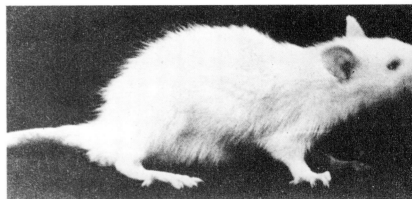

7.3 What is in our food?

You can now find out whether common foodstuffs contain sugar, starch, fat, or protein. Carry out these tests on a variety of foodstuffs and record your results as suggested in *table 13*.

Table 13
A suitable way of recording tests on food. You can use ticks to show substances present and crosses to show those absent (or not detected).

Food	Starch	Reducing sugar	Fat	Protein	Investigator's initials
bread					
carrot					
ham					
etc.					

1   *Starch* Add a few drops of dilute iodine solution directly to the surface of the food. If you see no change, put a little of the food material in a mortar and grind it with a little water. Then test the suspension.
2   *Reducing sugar* Grind a little of the food with water. Pour some of the resulting solution into a test-tube. Add 2 cm³ of Benedict's solution and bring to the boil. Boil in a beaker as before for a few seconds.

3 *Fat* Cut the food into thin shavings and soak a few of the shavings in ethanol to extract any fat. Warm the tube in your hand or a water bath to help this process. Allow the shavings to settle, then pour the clear ethanol into a test-tube almost filled with water.

4 *Protein* Grind a little of the food with water. Pour 2 cm³ of the resulting suspension into a test-tube. Add 2 cm³ of Millon's reagent and bring to the boil as before.

You can compare your results with the table in the Appendix (page 245).

7.4 Vitamins

The idea that a disease could be caused by the lack of an essential substance in food seems to have occurred to a number of people in the past, but only comparatively recently have carefully controlled experiments been done to test this idea. The first person to do such experiments was Magendie, who fed dogs on a diet of sugar, butter, gelatine (a protein), and distilled water. The dogs all died within 30 to 36 days.

The first disease shown to be caused by lack of an essential food substance was *scurvy*, which occurs when fresh fruit and vegetables are not included in the diet. Records of voyages of Vasco da Gama, Drake, Hawkins, and others from the fifteenth century onwards make frequent mention of the ravages of scurvy. Their ships were provisioned almost entirely with dried, pickled, and salted foodstuffs.

In 1753 a ship's doctor, James Lind, published *A treatise on the scurvy*, in which he described how sailors suffering from scurvy were cured within a week by eating two oranges and one lemon daily. During his voyages in 1772 and 1775, Captain Cook kept his ship's company free from scurvy by giving them plenty of fresh food. In 1804, an Admiralty order provided for a regular issue of lime juice to all sailors in the Navy. Merchant ships had to wait till 1853 for a similar regulation. The substance in fruit and vegetables which prevents scurvy was later called vitamin C. It was chemically isolated in 1928 and synthesized in 1933 and is now called *ascorbic acid*. It can be made cheaply and is used in tablet form on Polar and mountaineering expeditions. About 0.01 g per day is enough.

Ascorbic acid is a powerful reducing agent, easily oxidized in air, so it disappears from food left to stand and it is rapidly destroyed during cooking. It removes the colour from a blue dye called dichlorophenol indophenol (DCPIP for short), and we can use this property to estimate the amount of ascorbic acid in fruit juices.

Test for ascorbic acid

1 Place exactly 1 cm^3 of DCPIP solution in a test-tube using a syringe or a pipette. Do not shake the solution.

2 Fill a 1 cm^3 syringe with a 0.1 per cent solution of ascorbic acid.

3 Add the ascorbic acid solution to the DCPIP drop by drop, stirring gently with the syringe needle. Do not shake. Continue adding ascorbic acid solution until the moment when the colour has been removed from the DCPIP. Note the volume of ascorbic acid solution added.

4 Now repeat *1–3* using freshly squeezed lemon juice, boiled lemon juice, and any other liquid you want to test. In each case note the volume of juice required to remove the colour from the DCPIP solution.

5 Calculate the percentage of ascorbic acid in the juice as follows:

$$\text{percentage ascorbic acid in juice} = \frac{\text{volume of standard ascorbic acid solution added in } 3}{\text{volume of juice added in } 4} \times 0.1$$

Q1 Why should the solutions not be shaken during this test?

Q2 What effect did boiling have on the concentration of ascorbic acid in the lemon juice?

Other vitamins were discovered by chance observations of diseases caused by certain diets. In 1896 a doctor called Eijkman at a military hospital in Indonesia noticed that his hens, fed on the leavings from patients' meals (mostly 'polished' rice), had a paralysis similar to the human disease 'beri-beri'. A new hospital director stopped him using this food so Eijkman fed his hens on cheap unpolished rice and was surprised to find that they recovered. Later, the vitamin involved (now called *thiamine*) was extracted from rice husks. Thiamine is one of several substances which come under the general heading of vitamin B.

At the beginning of this century Frederick Gowland Hopkins, working in Cambridge, carried out a series of experiments similar to those of Magendie. He fed young rats on a diet of purified casein (milk protein), starch, sucrose, lard, and inorganic salts. He found that the rats did not grow and soon died. Adding 3 cm^3 of milk each day to the diet caused the rats to grow normally and stay healthy. The results of one of his experiments is shown in *figure 97*.

Q3 What does this tell us that Magendie's experiment with dogs does not?

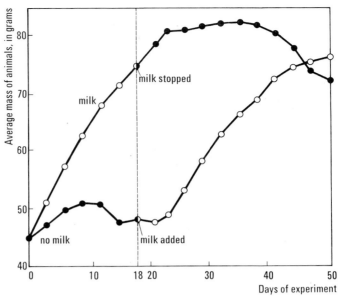

Figure 97
Frederick Gowland Hopkins's experiment on feeding milk to rats.

Graph labels: Average mass of animals, in grams (y-axis); Days of experiment (x-axis)

milk stopped

milk

no milk

milk added

Q4 In what other ways did Hopkins show a better knowledge of experimental procedure than Magendie?

Q5 Can you suggest a reason why the very young rats without milk grew a little during the first few days of the experiment?

Q6 Hopkins studied the average mass of the rats. What other measurements or observations might be taken as a record of normal healthy growth?

We now know that milk contains vitamins A and B with smaller amounts of others. This is why it helps animals to grow and stay healthy. We also know much more about the chemical nature of vitamins and what they do in our bodies. Most of them are used to make chemicals needed in small amounts for vital processes (*e.g.* cell respiration). Vitamin A is used to make a substance which is present in our eyes and enables them to work in dim light. If vitamin D is missing from the diet of growing children, the bones do not form properly, and the mass of the body may cause the legs to distort (see *figure 98*).

This condition, called *rickets*, was so common in industrial areas of Britain during the last century that it became known as 'the English disease'. Fish store vitamin D in their livers, so oil from these is a rich source of the vitamin. Giving children plenty of sunlight also helps in preventing rickets; alternatively, since only the invisible ultra-violet rays from the sun are effective, radiation from ultra-violet lamps can be used. The ultra-violet rays convert a substance in the skin into vitamin D.

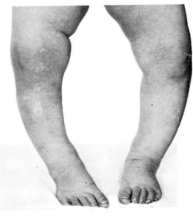

Figure 98
Rickets: one result of a deficiency of vitamin D in man.

Q7 Can you explain why rickets rarely occurs in children who live in rural areas?

By now you will have seen that a number of substances come under the general heading of vitamins. Details of some of these are shown in *table 14*.

Vitamin	Function in body	Effect of deficiency	Good food sources
A retinol	aids night vision, assists growth, protects eye surfaces and lining of breathing system.	night blindness, poor growth	fish-liver oils, liver, butter, and margarine
B	all vitamins in this group are concerned with the chemical process of respiration, and are also involved in other chemical changes in the body		all occur in liver and yeast, cereals and cereal products, meat, and fish
	1 (B$_1$) thiamine	'beri-beri' (nervous paralysis and muscular weakness)	
	2 (B$_2$) riboflavin	growth checked, skin becomes dry	
	3 nicotinic acid	pellagra (skin dry and red, growth checked, digestive disorders)	
C ascorbic acid	formation of connective tissue	growth checked; absence causes scurvy	fresh fruit and vegetables
D calciferol	absorption and use of Ca and P for bone and tooth building	rickets, poor teeth	fish-liver oils, fatty fish, dairy products; made in skin by action of ultra-violet lights

Table 14

7.5 Mineral elements

The food substances so far described are almost all made from the elements carbon, hydrogen, oxygen, nitrogen, and sulphur. Many other elements are present in our bodies and have to be obtained from the food we eat. Some elements occur in quite large amounts while other important ones are present in much smaller quantities. Details of other elements are given in *table 15*.

Element	Amount in body of man of mass 70 kg	Found in, or used for	Good food sources
calcium	1.0 kg	bones, teeth, body fluids	milk, cheese, bread, flour, green vegetables
phosphorus	700 g	bones, teeth, ATP, some proteins (*e.g.* casein in milk)	nearly all foods
chlorine	100 g	body fluids	common salt, and most foods
sodium	100 g	body fluids	common salt, and most foods
potassium	250 g	inside cells	most foods
magnesium	25 g	in bones, and in cells	plant foods
iron	3 g	red blood cells	liver, beef, some vegetables
iodine	0.03 g	thyroid gland	drinking water, fish, iodized table salt

Table 15

7.6 Water

A man can survive for a month without food, but without water he can only live for a few days. Water makes up about 70 per cent of the body's mass and we must take enough each day to balance the water lost from the body. (See *figure 99* and *figure 210*, page 230.) If you lose a lot of water by sweating, you must drink more to replace it; if you drink a lot, the surplus water is removed in the urine.

Figure 99
The daily water balance.

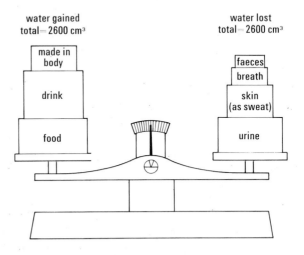

water gained total = 2600 cm³

made in body

drink

food

water lost total = 2600 cm³

faeces

breath

skin (as sweat)

urine

7.7 Energy requirements

While you are asleep, you use energy for breathing, for the heart beat, for keeping warm, and for other bodily functions. While awake you use all this, plus extra energy for physical work. The total amount used by an individual depends on his size and activity. The energy requirements of a number of individuals are shown in *figure 100*.

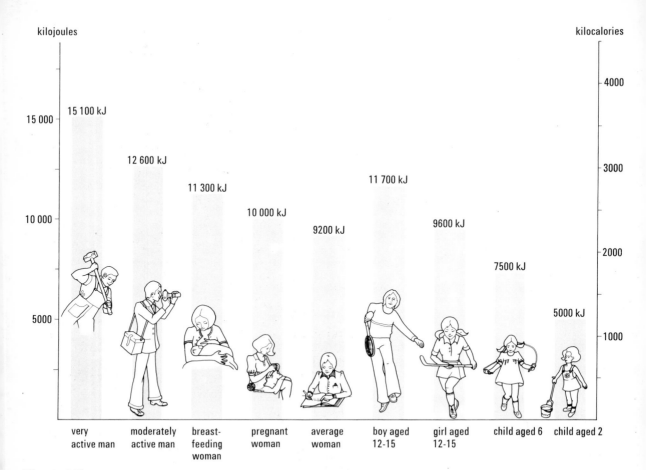

kilojoules

15 000

15 100 kJ

12 600 kJ

11 300 kJ

10 000 kJ

9200 kJ

11 700 kJ

9600 kJ

7500 kJ

5000 kJ

10 000

5000

kilocalories

4000

3000

2000

1000

| very active man | moderately active man | breast-feeding woman | pregnant woman | average woman | boy aged 12-15 | girl aged 12-15 | child aged 6 | child aged 2 |

Figure 100
The daily energy requirements of various people.

Other nutritional requirements of some of these individuals are shown in *table 16*.

		Boy 12–15	Girl 12–15	Man (moderately active)	Woman (most occupations)
Energy	kcal	2800	2300	3000	2200
	kJ	11 700	9600	11 600	9200
protein	minimum g	45	44	45	37
	recommended	70	58	75	55
calcium	mg	700	690	510	510
iron	mg	14	14	10	12
Vitamin A	mg	730	740	750	750
thiamine	mg	1.1	0.9	1.2	0.9
riboflavin	mg	1.4	1.4	1.7	1.3
nicotinic acid	mg	16	16	18	15
ascorbic acid	mg	25	25	30	31
vitamin D	mg	2.5	2.5	2.4	2.4

Table 16
Daily requirements of individuals in energy and in nutrition. *Data from Ministry of Agriculture, Fisheries and Food (1970)* Manual of nutrition *(7th edition) H.M.S.O*

Study *figure 100* and *table 16* and then try to answer the following questions.

Living things in action

Q1 Why do women have a lower energy requirement than men who have similar occupations?

Q2 Why does a woman's calcium requirement increase when she is *a* pregnant and *b* breast-feeding a baby?

Q3 Why does pregnancy increase a woman's energy requirements?

Q4 Why are the protein requirements of young teenagers greater than those of adults, even though the adults are greater in body mass?

Q5 Why do women and teenage girls have a greater need for iron than men and teenage boys?

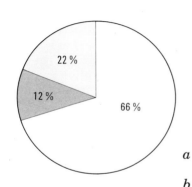

energy intake per day:

☐ over 11 300 kJ (over 2700 kcal)

▨ 9200-11 300 kJ (2200-2700 kcal)

☐ under 9200 kJ (under 2200 kcal)

Figure 101
A pie chart showing the distribution of world population according to average daily intake of energy.

The world's food problem

From time to time disasters such as floods, earthquakes, wars, droughts, or the failure of a major crop may cause severe *famine* in a particular area. This is often followed by a widely publicized rescue operation to relieve the situation. *Malnutrition,* however, does not so often make the headlines; but it is widespread and in many places continuous.

Malnutrition is caused by a diet which in some way does not supply what a person needs. It takes many forms:

a Two-thirds of the world's population receive barely enough food to supply the energy they need. (See *figure 101*.)

b In Indonesia children may go blind through vitamin A deficiency coupled with inadequate medical services.

c In Central America children frequently suffer from pellagra, caused by an imbalance of vitamins and amino acids. This, in turn, is caused by a diet consisting of little else but maize.

d In Britain's industrial cities rickets used to be very common because of diets low in vitamin D and insufficient exposure to sunlight. Rickets still occurs in some British cities.

e A lack of sufficient protein in the diet is common in some deprived areas of the world and gives rise to the condition known as *kwashiorkor* (see *figure* 90).

f Wealthier people all over the world tend to eat too much and become overweight.

If some people in a community are so obviously badly nourished that a doctor can diagnose this without doing any laboratory tests, then there must be others who are also getting poor diets without appearing ill. Malnutrition occurs most often among children aged between six months and five years. They are growing fast and need diets of good quality; yet because they are small and dependent on others to feed them, they may not get enough food. Young children also pick up infections very easily. Illness and malnutrition often make each other worse. Suppose an underfed toddler lives in an unhygienic slum house and catches an infectious disease from which he dies. We cannot say for certain whether it was the illness or the malnutrition which killed him. But we do know that *both* contribute to the high death rates among children in developing countries. A family with many children is often the one where malnutrition is the worst problem. This is one reason why family planning and the spacing of children are advocated in many of these countries.

In studying a country's food problems, an expert on nutrition must find out a lot about the country and its people. He must find out what foods are grown, whether they are scarce at some times of the year, what foods the people like best, how the food is cooked, and what foods are given to children. He must know what foods are imported and how easily food can be transported across the country. He must compare the cost of food with the income of the lowest paid. He must ask how many times a day people have meals, whether the children get meals at school, whether the mothers of young children get any extra provision through baby clinics and whether such clinics give advice on feeding babies. Then he would also need to know something about the wealth of the country – its national income, its welfare and health policies, and how the wealth is distributed between rich and poor, town and country, and one region and another. He would also need to look at the ways in which malnutrition shows itself in the country.

Only then will he have some idea of how and why malnutrition occurs. There is really no *one* great 'world food problem' – each country has its own problems. Although trade and diplomatic relations exist between countries, the countries themselves still differ enormously. Indeed, in a large country, like India or Nigeria, the people of different regions eat quite different foods and may have to be considered separately.

All the same, when we are looking at a country or region, we can make a good guess at the kind of people who will be malnourished. They are most often young children – especially those of the poorer families. *Figure 102* shows that in Brazil the richer people have a higher energy intake. The poor families of Brazil have a much worse diet than the poor families of Britain, because food is cheap for

Plate 1 (page 33)
A missel thrush *(Turdus viscivorus)*. How is the shape of a bird's body adapted to overcoming wind resistance?
Photograph, Eric Hosking.

Plate 2 (pages 41, 53, 194), below.
a An Eskimo in wolfskin clothing trimmed with caribou, in the North West Territories of Canada.
b A bushman boy in Botswana. Both **a** and **b** illustrate how the shape of the body is adapted to the control of body temperature in a particular environment. They also show how man deliberately adapts to his environment by his dress.

a   b

Photographs: **a,** *Popperfoto;*
b, *Terence Spencer/Colorific!*

Plate 3 (pages 67 and 76), right.
A corrosion preparation of the
human lungs.
*Photograph reproduced by kind
permission of the President and
Council of the Royal College of
Surgeons of England.*

Plate 4 (page 87),
background and below right.
Luminescent organisms.
a Glow-worm Grotto in Waitomo
Caves, New Zealand.
b A luminous toadstool, in
Malaysia.
*Photographs: **a**, New Zealand
High Commission; **b**, Ivan
Polunin/Natural History
Photographic Agency.*

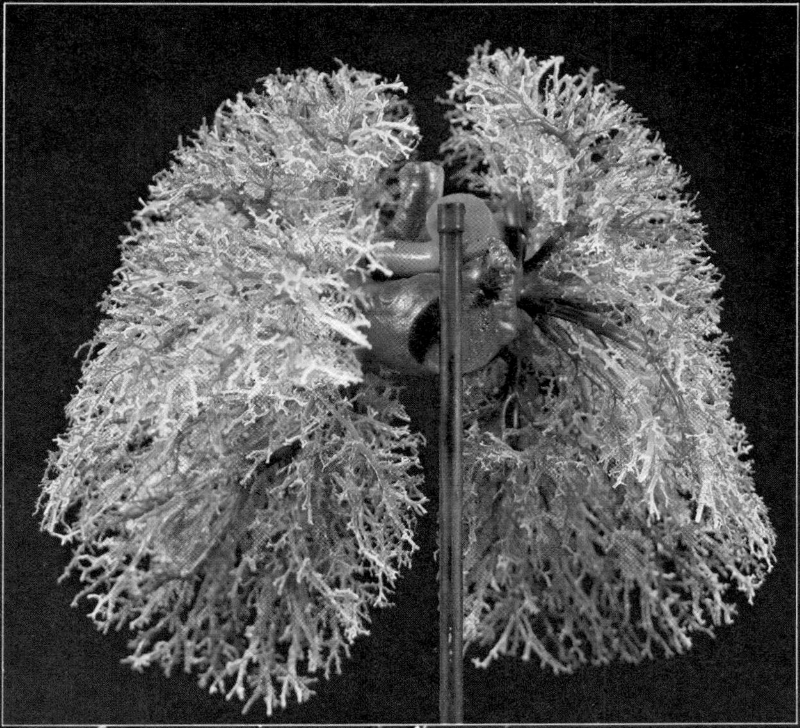

a

b

c

Plate 5 (page 130), above.
Some ways in which
animals obtain their food.
a The tentacles of a dahlia
anemone *(Tealia felina)* engulfing
a blenny.
b A spider tying up a bee before
eating it.
c Goose barnacle *(Lepas
fascicularis)* feeding underwater.
*Photographs: **a** and **c**, Heather
Angel;* **b**, *Treat Davidson/Frank
W. Lane*

Plate 6 (page 166)
Sunlight streaming through beech
leaves.
Photograph, Heather Angel.

Plate 7 (page 166), below, top.
The spectrum obtained from white
light.

Plate 8 (page 167), below, bottom.
The absorption spectrum of
chlorophyll, that is, the spectrum
obtained by passing white light
through an extract of chlorophyll.

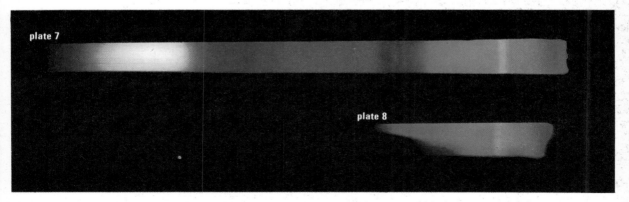
plate 7

plate 8

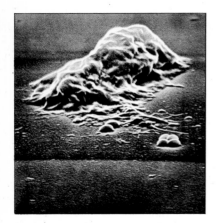

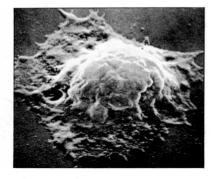

Plate 9 (page 216)
Three stages in the consumption
of staphylococcus bacteria by a
large white cell in human blood.
(× 5000 approximately.)
From Nilsson, L. (1974) Behold
man, *Harrap.*

Plate 10 (page 218)
A simplified diagram of blood
circulation in man.

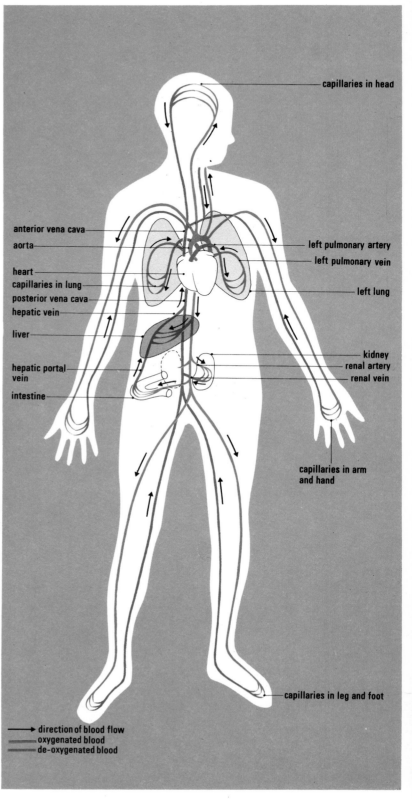

capillaries in head

anterior vena cava
aorta

left pulmonary artery
left pulmonary vein

heart
capillaries in lung
posterior vena cava
hepatic vein

left lung

liver

kidney
renal artery
renal vein

hepatic portal
vein

intestine

capillaries in arm
and hand

capillaries in leg and foot

→ direction of blood flow
oxygenated blood
de-oxygenated blood

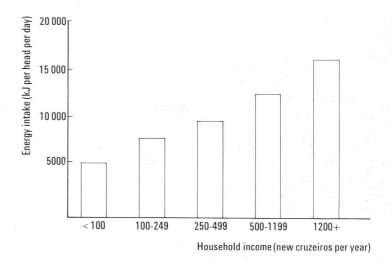

Figure 102
A bar chart showing the intake of energy of Brazilian families in relation to various incomes.

the lowest paid workers in Britain compared with those in Brazil. Also, Britain has various welfare and social service benefits, such as unemployment pay, so that the poorest British families are always better off than the poorest Brazilians.

Not only is the amount of money in a country important. How it is distributed is important too. If a country is becoming richer, its wealth must be spread out among all the people and not concentrated among the rich. Thus, some form of distribution of wealth is essential, as well as economic development itself. Development need not mean industrialization. It could mean an increase in farm production and the export of crops, including surplus food.

People still argue about the best way of dealing with malnutrition. Should there be more doctors and hospitals or more agricultural schools? Should there be more education about nutrition in the schools or better birth control policies to reduce the size of families? Should there be more political action to attempt to achieve a better distribution of land and wealth? How can international aid be made more effective so that the contrast in diets illustrated overleaf in *figure 103* is made less pronounced?

More hospitals may benefit people in towns, but not those living far away from a main road. Better food production is of little use if transport and marketing systems cannot handle it efficiently. Increased food production often means using more fertilizers and pesticides which are expensive for farmers who have low reserves of cash. Education in schools may conflict with what children learn from their parents at home. Political changes may only result in the enrichment of a new group of politicians. It may be simply impossible to distribute resources fairly. A programme which works perfectly in Chile might fail in Ghana and vice versa. We must find out what people eat – and why – before trying to persuade them to do differently. We should find out, if possible, what stops people from getting a satisfactory diet. Then we might be able to find ways of improving the situation in a manner which suits the way of life of the country concerned.

Figure 103
The daily diet in an American city and in India.
Based on data from FAO and USDA.

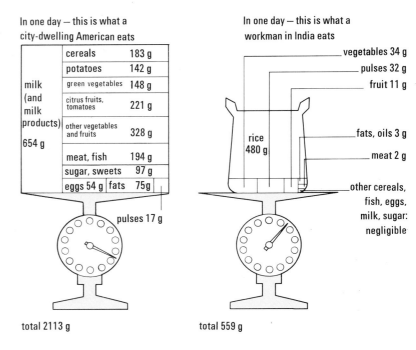

In one day — this is what a city-dwelling American eats

milk (and milk products) 654 g	cereals	183 g
	potatoes	142 g
	green vegetables	148 g
	citrus fruits, tomatoes	221 g
	other vegetables and fruits	328 g
	meat, fish	194 g
	sugar, sweets	97 g
	eggs 54 g	fats 75g

pulses 17 g

total 2113 g

In one day — this is what a workman in India eats

vegetables 34 g
pulses 32 g
fruit 11 g

rice 480 g

fats, oils 3 g
meat 2 g

other cereals, fish, eggs, milk, sugar: negligible

total 559 g

How animals feed

8.1 Feeding in man

Some of the food you eat, such as soup or ice cream, can be swallowed straight away. Harder foods have to be broken up first. You break them up by biting and chewing with your teeth, which are set in your jaws. (See *figure 104*.) The upper jaw is fixed to the rest of the skull while the lower jaw can be moved by muscles in the cheeks.

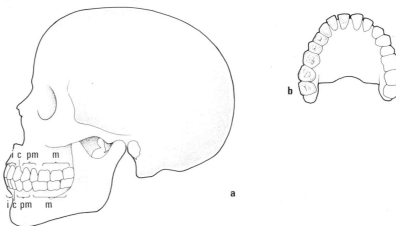

Figure 104
a Side view of human skull
i = incisors, c = canines,
pm = premolars, m = molars.
b Upper teeth seen from below.

As you chew, the food is moistened with saliva and moved about by the tongue. This moistening enables you to taste your food as well as making swallowing and chewing easier. Only when chewing (*mastication*) is complete, is the food moved to the back of the mouth and swallowed.

You can find out more about human teeth in several ways (*see below*). In each case make a note, with drawings if necessary, of the relative size and shape of the teeth.
1   Use a mirror to examine your own teeth.
2   Look at the teeth of other members of your class or family.
3   Examine the teeth in a human skull or a model of one.
4   Look at casts of jaws and teeth made by dentists.
5   Examine teeth which have been extracted by dentists. Such specimens are likely to be decayed and may also have fillings.

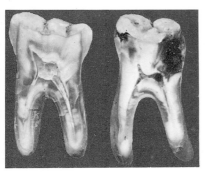

Figure 105
Human molar teeth
cut in half vertically
(× 2, approximately).
The righthand tooth shows
dental caries.

The arrangement of an animal's teeth is called its
dentition. Most mammals have four types of teeth.
Nearest the front of the mouth are the *incisors.* Behind
these lie the *canines* and cheek teeth. In fact the cheek
teeth are of two types – *premolars* nearer the front and
molars at the back. These appear to be similar in many
mammals, but are distinguished by the way in which they
develop. In young mammals a first set of deciduous or milk
teeth is gradually replaced by a permanent set. Molars do
not replace teeth in the deciduous set, but appear for the
first time in the permanent dentition.

Q1   From the observations you have made, can you suggest
different functions for these four types of teeth?

Q2   How does your own set of teeth differ from the full
permanent dentition illustrated in *figure 104*?

Q3   How much variation is there among the dentitions of the
different members of your class?

Q4   Try to find out what wisdom teeth are. Where and when do
they first appear?

8.11 Tooth structure

Figure 105 shows photographs of human molar teeth which
have been sawn in half and whose cut surfaces have been
ground and polished. Compare the photographs with the
diagram in *figure 106*.

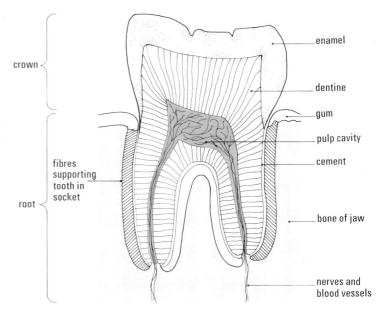

Figure 106
A diagram of a vertical section of
one human molar tooth.

The *enamel* is particularly hard and resistant to decay. It is one of the hardest natural substances and for this reason is sometimes the only remaining part of certain fossil animals. *Dentine* is softer and, unlike the enamel, contains living tissue. The fine canals which radiate outwards from the *pulp cavity* contain fibres from living cells which manufacture the dentine around them. In the photographs the pulp cavity appears to be empty. In the living tooth, however, it houses blood vessels, which supply the cells of the tooth, and also nerves. (See *figure 107*.) The roots of the teeth are covered with a thin layer of *cement* from which run elastic fibres which support the teeth in its socket and act as shock absorbers.

8.12 Keeping our teeth

Have you got perfect teeth? If so, you are both fortunate and unusual. One-third of all the people in this country over the age of 15 have lost all their natural teeth. Three-quarters of children aged five have some decayed teeth. Yet most people living in what we consider to be primitive societies have perfect teeth.

Much of the carbohydrate-rich food that you eat, such as bread, cake, biscuits, sweets, and cooked potato, is soft and tends to cling to the teeth and collect in the spaces between.

Bacteria in the mouth break this material down and produce acids which dissolve the tooth enamel. Though saliva helps in rinsing the mouth, there are substances in it which get deposited on the tooth surfaces. This film of *tartar* absorbs the acids formed by the bacteria and brings them into direct contact with the enamel. Once the enamel has been penetrated, the acid breaks down the dentine. Bacteria themselves enter the dentine and the resulting decay forms a cavity which enlarges till it reaches the pulp cavity – then you feel the pain.

To get and keep good teeth you must do certain things and avoid others. First, your diet must include the materials needed for making teeth. If you look in Chapter 7 you will find out what these food substances are. Clearly you cannot avoid eating all soft foods, but you can make sure that as little food as possible remains in the mouth afterwards. One way of doing this is to finish the meal with something crisp, such as an apple or celery; another is to rinse the mouth with water. Brushing the teeth, with or without toothpaste, is even more effective – if it is done correctly. The teeth should be brushed away from the gums towards the biting surfaces so that food particles are removed from the spaces between the teeth. (See *figure 108*.) It is also just as important to brush the inner as the outer surfaces of the

Figure 107
A thick section through part of the lower jaw of a kitten (× 60, approximately).

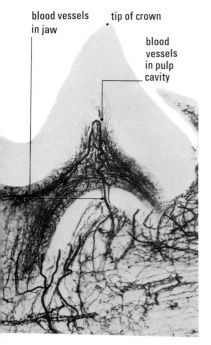

blood vessels in jaw   tip of crown

blood vessels in pulp cavity

teeth. Proper brushing takes two or three minutes. Ideally it should be done after every meal. It should certainly be done after breakfast and last thing at night. After brushing, the mouth should be thoroughly rinsed with water. Brushing also benefits the gums, which are diseased in many people.

Most people like sweets and sweet foods, but these are the foods which most readily produce acid and cause decay. Few things are more harmful to teeth than sucking ice 'lollies' or other sugary materials which keep sugar in contact with the teeth for a long time.

Finally, see that your teeth are regularly inspected by a dentist. He can spot early signs of decay or gum disease and deal with them before they get too bad.

Figure 108
How to clean your teeth efficiently. *By courtesy of the British Dental Association.*

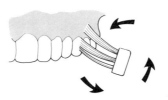

a *Brush down on the upper teeth*

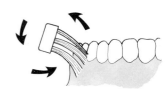

b *Brush up on the lower teeth*

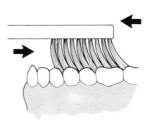

c *Brush the chewing surfaces with a scrubbing stroke*

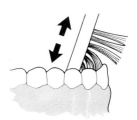

d *Brush behind the front teeth using an up-and-down stroke with the length of the brush head*

8.2 Feeding in other mammals

If possible, watch a cat feeding and look at a film loop which shows extra details of how it does so. Then examine a cat's skull and study *figure 109*.

Now try to answer the following questions:

Q1 Can you identify the same types of teeth in a cat as you saw in your own mouth?

Q2 Which of the cat's teeth appear particularly well developed? Why do you think this is?

The photographs in figures 109 to 111 are reproduced by courtesy of the Wellcome Museum of Medical Science.

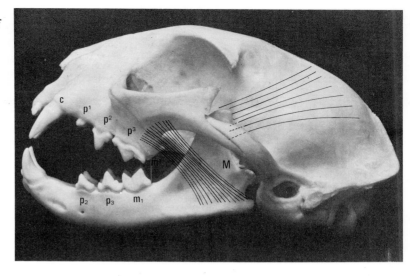

Figure 109
The skull of an adult domestic cat in side view (× ⅘).

Q3　When a cat's mouth closes, do the cheek teeth meet crown to crown? What effect will this have on the food within the mouth?

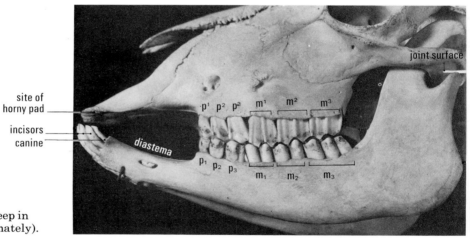

Figure 110
The skull of an adult sheep in side view (× ½, approximately).

Now look at a film loop of plant eaters (*herbivores*) eating and examine a sheep's skull. Study *figure 110* and then try to answer the following questions:

Q4　Which type of teeth does the sheep lack?

Q5　How do you think the sheep manages to bite off its food?

Q6　Do the cheek teeth meet when the jaw closes?

Q7　How do the surfaces of the cheek teeth (see *figure 111*) differ from those of the cat? Can you suggest a reason for this?

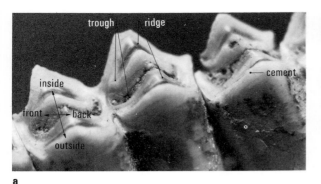

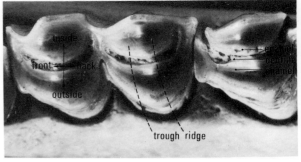

a   **b**

Figure 111
Details of the crowns of
cheek teeth from the left
lower jaw of the sheep's skull
in *figure 110*. **a** in side view,
b seen from above.
(× 4, approximately)

Q8 In what way does the arrangement of dentine and enamel
in the cheek teeth help in the chewing process? (*Figure 112.*)

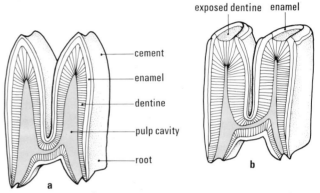

Figure 112
Stereograms of herbivore cheek
teeth.
a newly erupted, **b** after wear.
After Mackean, D. G. (1973)
Introduction to biology,
5th edition, John Murray.

Now compare the lower jaws of both mammals which are
illustrated in *figure 113*. The temporal muscle runs from the
jaw to the side of the skull. When it contracts, the jaw is
closed with a scissor-like action, pivoting on the fulcrum.
The masseter muscle runs from the jaw to the cheek and
pulls the jaw forward as well as closing the mouth.

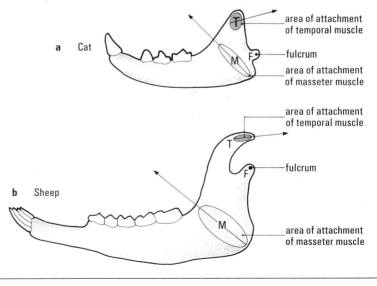

Figure 113
The lower jaws of cat and sheep,
showing the attachment of
temporal and masseter muscles.
The arrows indicate the direction
in which the contracting muscles
pull.

	Cat	Sheep
Largest of the two muscles	temporal	masseter
Distance of line M from F (mm)		
Distance of line T from F (mm)		

Table 17

Use the drawings, or real jaws, to complete *table 17*. Then use the completed table to answer the following questions. Models of the jaws may also help you.

Q9 Which of the muscles attached to the cat's jaw will exert the greatest leverage about the fulcrum?

Q10 How does your answer to question 9 fit in with your own observations of a cat feeding?

Q11 Which of the muscles attached to the sheep's jaw will exert the greatest leverage about the fulcrum?

Q12 What feature must the joint, where the jaw is attached to the skull, possess to enable the sheep to make the kind of jaw movements you have observed?

Look again at a sheep's skull and jaw and see if the joint appears to have the feature you have just suggested.

8.21 Dentition and diet

You will realize from your studies that the dentition of a mammal is well adapted to its diet. The teeth and the jaws look as if they were 'made for the job' since their structure is closely allied to their function. Of course, merely recognizing such an adaptation is not the same as explaining how it came about; nor is it the whole story. Other features are also related to diet.

The behaviour of a meat eater (*carnivore*) such as the cat includes actions like stalking and catching prey. The cat's sharp eyes and ears, well developed sense of smell, powerful body, and sharp claws all help in this. The carnivore's digestive system is also well adapted to its diet. The behaviour patterns of plant eaters (*herbivores*) are also adapted to their way of life. They tend to live in flocks or herds where their large numbers give them some protection from would-be attackers. Their eyes are usually positioned so that they have a good all round view and in many cases they are capable of running away from predators for quite long periods. Their digestive systems are also adapted to their diet – in this case, grass. No part of an animal's body should be studied in isolation from the rest, except as a starting point for learning more about it.

Figure 114
The head and part of the rest of the body of a desert locust ($\times$ 40).
Photograph, Shell.

8.3 Feeding in insects

An expert on insects once wrote:
'Almost every kind of substance of plant or animal origin serves to nourish some insect or other . . . from fluids such as blood and plant sap to the hardwood of trees, bones and other seemingly indigestible substances . . . it is not surprising that with so diversified a menu the food-getting mechanisms of insects display an extraordinary range of devices.'

A locust (see *figure 114*) is estimated to consume a mass of food equal to its own mass, per day. Since locusts live in swarms of 1 000 000 000 insects or more, maintaining such a swarm uses up 3000 tonnes of plant material every day. The feeding problem facing a locust resembles that of a sheep, but it is adapted to its feeding in a very different way.

Q1   What is this problem in feeding which faces both the sheep and the locust?

You can study the feeding action of a locust by placing one in a small plastic box with a few blades of grass. You may also be able to see a film loop of a locust feeding. Then examine the mouthparts of a dead locust.

1   Pin the locust on its back under a dissecting microscope or hand lens, with its head pointing away from you.
2   Try and distinguish the mouthparts listed below. Use a mounted needle to separate them if necessary.
a   A large upper lip (*labrum*), which hides most of the other mouthparts. If you fold this back, the rest will be exposed.
b   A pair of relatively huge black jaws (*mandibles*), which are rocked towards and away from each other by powerful muscles (see *figure 115*).

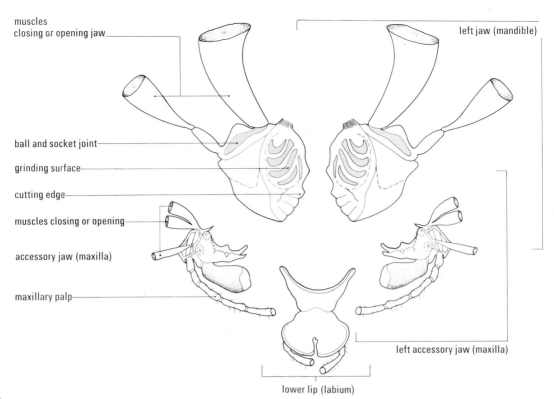

muscles closing or opening jaw

left jaw (mandible)

ball and socket joint

grinding surface

cutting edge

muscles closing or opening

accessory jaw (maxilla)

maxillary palp

left accessory jaw (maxilla)

lower lip (labium)

Figure 115
The mouthparts of a locust (partly diagrammatic) removed from the head. Note the large mandibles with their grinding and cutting surfaces.

c A pair of small additional jaws (*maxillae*), which lie just below the mandibles. Each consists of three parts. The innermost of these is jaw-like and assists the mandibles in cutting vegetation and pushing it into the mouth.

d A shovel-like lower lip (*labium*), at the floor of the mouth.

Figure 116
A sectional view of a locust's head as seen from the animal's right. This gives an idea of the relationship of the mouthparts to each other.
After Thomas, J. G. (1963)
Dissection of the locust, *Witherby*.

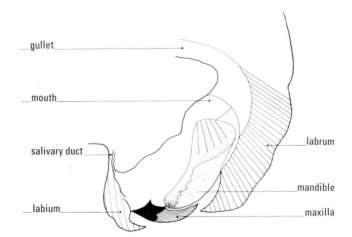

gullet

mouth

salivary duct

labium

labrum

mandible

maxilla

While the food is being chewed, saliva is poured onto it through a duct which opens into the mouth. Even after it is swallowed the food is further ground up in the *gizzard* – a muscular organ with a horny lining of ridges and spines.

Q2 Do the cutting edges of the mandibles overlap when they are closed?

Q3 Are there any signs of wear on these edges?

Q4 Examine the jointed structures (*palps*) which are attached to the maxillae and labium. What might their function be? Your observations of the animal feeding may provide evidence here.

The housefly feeds in a very different way from the locust. (See *figure 117*.) You can study this in more detail.

Figure 117
A housefly feeding on bread and honey. Note the extended proboscis.
Photograph, Heather Angel.

1 Examine the head of a freshly killed housefly. Look at the *proboscis* (mouthparts) with a lens or binocular microscope so that you can see its general structure and its relationship to the head.
2 Watch the way a housefly feeds. You can either do this by imprisoning a fly in a small plastic container with a drop of sugar solution or by watching a film loop.
3 Examine the photomicrograph of a housefly proboscis in *figure 118*.

Q5 In what ways do a housefly's mouthparts differ from those of a locust?

Sections

at A

labrum

Figure 118
The mouthparts of a housefly.
a Attached to part of the head
skeleton (× 100).
b The rectangular area of the
fleshy lobes, shown in more detail
(× 1500).

at B

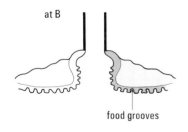

food grooves

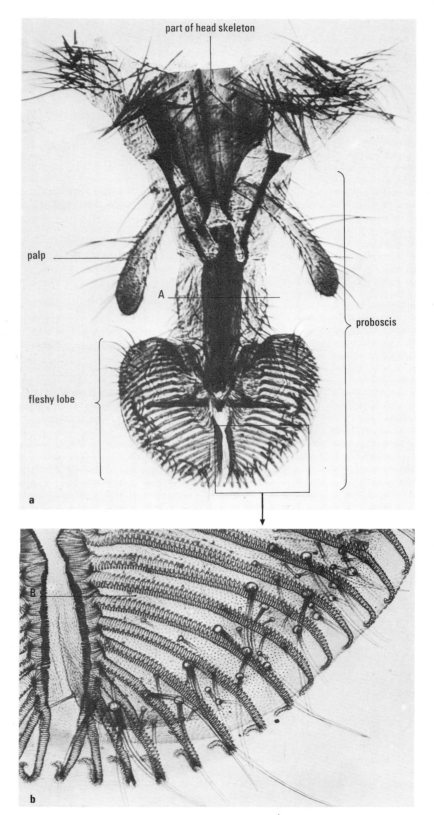

part of head skeleton

palp

A

proboscis

fleshy lobe

a

B

b

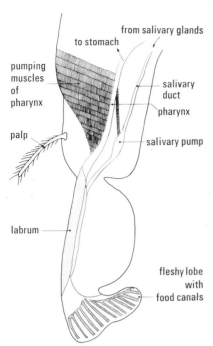

Figure 119
A diagrammatic section through
the proboscis of a housefly.
*Based on Borrodale, L. A., and
Potts, F. A. (1961)* The
Invertebrata *4th edition, revised
by Kerkut, G. A.; Cambridge
University Press.*

The portion of the proboscis attached to the head contains
two tubes; a narrow one down which digestive juices
(saliva) pass, and a wider one up which food passes. In the
fleshy lobes which form the end of the proboscis these
tubes branch and spread out to form a series of parallel
food grooves (see *figure 119*).

When a fly alights on food it extends its proboscis by
muscular action and blood pressure, so that the lobes are
pressed into contact with the food. Saliva flows down the
salivary duct into the food canals and onto the food,
moistening it and dissolving it. By the action of the powerful
pharyngeal muscles shown in *figure 119*, the semi-liquid
food is sucked up into the pharynx and so passes into the
stomach. Normally only small solid particles (up to 6 μm
in diameter) can be taken in, but by separating the fleshy
lobes, flies can take in larger particles, including the eggs
of parasitic worms, in suspension in liquid.

Locusts may eat a lot of man's food, but the housefly does
just as much damage of another kind. The eggs of houseflies
are laid in decaying vegetables, dead animals, and animal
faeces, all of which provide food for the larvae which hatch
from the eggs. In hot weather, eggs may develop into adults
in as little as seven days. Flies feed on any available
organic matter, from refuse, faeces, and manure to human
food.

Q6   From your knowledge of the feeding mechanism of
houseflies, how do you think they might be involved in
spreading diseases? (See *figure 120.*)

Q7   Are there any other features of the housefly which might be
responsible for transmitting diseases?

Q8   Can you suggest ways in which the transmission of
diseases by houseflies might be reduced or even prevented?

8.4 Other ways of feeding

In this chapter you have studied your own teeth and you
have compared the way in which two other contrasting
mammals feed. You have also examined the feeding
mechanisms of two insects. The variety of ways in which
animals obtain and eat their food is enormous and a full
study of it could take a very long time. However, you could
get an idea of *some* of the mechanisms involved by finding
out answers to some of the questions which follow. In a few
cases you may be able to make direct observations of the
animals concerned, but you will certainly have to make use
of library books too. Colour *plates 5a, b,* and *c* may also
help you.

Q1 Whales, such as the blue whale, are amongst the largest animals which have ever lived, yet their food consists of relatively small animals in the plankton. How do whales obtain this food?

Q2 Most animals move about to find their food, but some, such as the sea anemone, spend most of their lives in one place. How do they catch and eat the food which comes to them?

Q3 Many snails, like sheep, feed upon green plant material; but the way in which they eat their food is very different. How do they do this?

Q4 Most spiders spin webs with which they catch small flying insects. How do they then kill their prey and obtain any food material from them?

Q5 At first sight the soil might not seem a promising source of food, yet it is inhabited by a very large number of earthworms and other animals. What do these earthworms feed on and how do they eat it?

Q6 You may have seen rose bushes infested with aphids (greenfly). Unlike the locust, these insects do not chew the leaves of the plants. What do they feed on and how do they get their food from the plant?

Q7 Barnacles have been described as 'standing on their heads, kicking their food into their mouths'. How do they extract their food from the water in which they live?

Background reading

Figure 120
A culture of bacteria from the tracks made by a fly.
Photograph, Radio Times Hulton Picture Library.

'It must be something in the water'

You will recall from section 8.12 that dental decay is caused by eating soft foods, especially sugary ones, and failing to clean the teeth afterwards. But other factors affect dental health. One interesting discovery was made in the United States; in some regions people tended to develop teeth, which, instead of being white, were mottled brown. They were nonetheless almost free from dental caries (decay). It was discovered that the drinking water in these areas contained unusually high concentrations of fluoride ions – up to 8 parts per million (p.p.m.). In other parts of the country, the amount of fluoride in drinking water varied from 8 p.p.m. down to the merest trace. At a level of 1 p.p.m. there was no mottling of teeth and little decay. Below 1 p.p.m. of fluoride, the teeth examined were found to be affected more and more by dental caries. The graph in *figure 121* shows the results of examining over seven thousand American school children between the ages of 12 and 14 from 21 different cities.

Figure 121
Correlation between the amount of dental caries and the fluoride content of drinking water.
From Dean et al. *(1942), quoted in the 'Memorandum on the dental health of children' (1959) of the British Dental Association.*

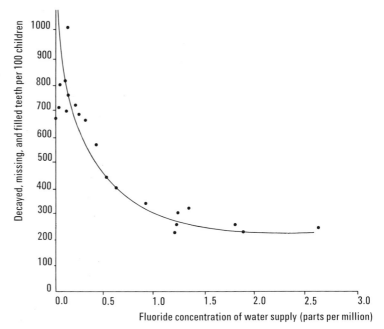

Figure 122
Proportion of children free from dental caries (in permanent teeth) in test areas in Britain.
From The Health Education Council (1970) Our teeth. Summary of the 1969 report on eleven years of fluoridation.

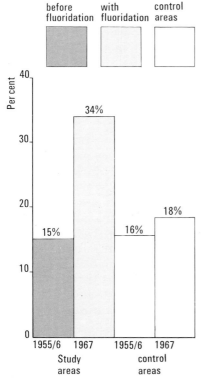

From data such as these, scientists concluded that it would be desirable to raise the fluoride level of drinking water to 1 p.p.m. if it was not naturally at this level already. Trials were carried out in a number of cities, with dramatic results. The level of dental caries fell by roughly 60 per cent when fluoride deficiency was made good in this way.

Fluoridation of British water supplies started between 1955 and 1956, when certain areas agreed to help in an experiment to test the effectiveness of fluoride in reducing dental decay. The water in Watford, part of Anglesey, and Kilmarnock was treated to bring the fluoride level up to 1 p.p.m. Three control areas (Sutton in Surrey, the rest of Anglesey, and Ayr) continued to receive unfluoridated water. After 11 years it was possible to compile results about children who had drunk fluoridated water all their lives (see *figure 122*); as expected, fluoride was causing an improvement in dental health. It was too early to say whether this would continue into later life, but in Canada and the United States, where some areas had received fluoridated water for 25 years, the results showed that older children benefited too (*figure 123*).

In case fluoride might have any unforeseen side effects, there was a careful study of the general health of people in control areas and fluoridated areas. To quote from a government report, 'During the eleven years under review, doctors reported only two patients with symptoms which they felt might have been associated with fluoridation. Careful investigation in both instances failed to attribute

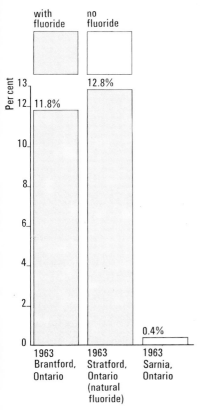

with fluoride | no fluoride

Per cent

13
12.8%
12 — 11.8%

10

8

6

4

2

0.4%
0

1963 Brantford, Ontario | 1963 Stratford, Ontario (natural fluoride) | 1963 Sarnia, Ontario

Figure 123
Proportion of 16- and 17-year-olds free from dental decay in test areas in Canada.
From The Health Education Council (1970) Our teeth. Summary of the 1969 report on eleven years of fluoridation.

the symptoms to the drinking of fluoridated water.' The same report concluded that '. . . the fluoridation of water supplied up to the level of 1 p.p.m. is a highly effective way of reducing dental decay and is completely safe'. Fluoridation is supported also by the World Health Organization, the British Medical Association, the Royal Society of Health, the General Dental Council, and the British Dental Association.

But not all parts of Britain use fluoridated water, since the local authority for each area has to decide whether or not to add fluoride. In fact, one of the three original trial areas voted to discontinue fluoridation before the 11 years had gone by. Perhaps you can find out which way *your* local health committee voted on the issue.

People oppose fluoridation for a number of reasons. Some of these can be discussed in scientific terms, but others reflect personal attitudes towards fluoridation. For instance, in one experimental area several people wrote to the Water Board complaining that the fluoridated water had an unpleasant taste; there had been some confusion about the date of fluoridation and the letters were in fact written *before* fluoride had been added to the water. Other anti-fluoride arguments include economic and moral issues.

Think about these objections:

1 Fluoride is wasteful. It costs over £2000 to provide every £1-worth of fluoride for children under 10; the rest goes down the drain or is drunk by adults who do not benefit from fluoride.

2 Fluoridation does not remove the *causes* of dental decay. Money spent on fluoridation would be better used in campaigns to encourage people to clean their teeth and to eat fewer sweets.

3 Fluoride is known to be poisonous. About 2500 mg is enough to kill a man.

4 Fluoridation infringes a person's right to freedom. Anyone wanting to take in fluoride can buy fluoride tablets or use fluoridated toothpaste. No local authority would force people to take tablets, yet putting fluoride into the public water supply compels everyone to consume it. Once 'mass medication' starts in this way, where would it stop?

Getting food to the body

9.1 Big molecules and small molecules

Once you have swallowed some food, you are no longer conscious of what happens to it, but you will probably know that it goes into your stomach. You may also know that this is part of a tube running through the body called the alimentary canal or gut. You may have seen in Chapter 6 that the energy for muscle contraction comes from food. But before your muscles can make use of the food which you eat, it has to pass through the wall of the gut and then be distributed to the muscles by the blood. The same must happen to the food required by any other part of the body. In section 7.22 you saw that starch molecules are long chains made up of glucose units. The glucose molecules themselves are therefore obviously very much smaller. The question now arises whether these molecules are able to pass through the gut wall and into the blood. This idea cannot very easily be investigated with the gut itself. However, you can use a model. Cellulose (Visking) tubing resembles the gut in the way that it allows molecules to pass through it.

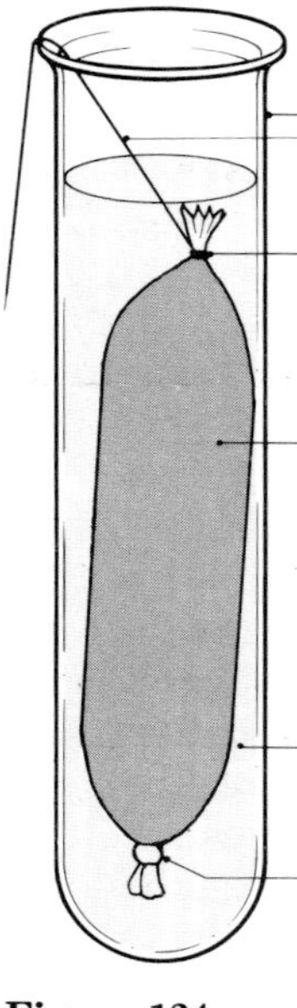

1 Close one end of a 15 cm length of Visking tubing by tying it tightly. Wet the other end so that the tubing can be opened up.

2 Using a syringe (without a needle) nearly fill the tubing with a mixture of glucose solution and starch suspension.

3 Tie a piece of thread round the top of the tubing, leaving a length with which you can handle it.

4 Rinse the outside of the tubing thoroughly under the tap.

5 Place the Visking tube 'bag' in a boiling-tube and fill the space between the Visking tubing and the glass with distilled water (see *figure 124*).

6 Immediately remove a small sample of the water in the boiling-tube and test it for *a* starch and *b* reducing sugar, using the same tests as in section 7.3 (see page 108).

7 Repeat these tests at intervals of approximately 10 minutes for the next 30–40 minutes. Make a careful note of

boiling-tube

thread for handling tubing

top of tubing sealed off by a knot

starch and glucose solution

water

leakproof knot tied in Visking tubing

Figure 124
Finding out whether glucose or starch can pass through Visking tubing.

your results each time. It is best to record them in the form of a table.

Q1 Why was the outside of the Visking tubing rinsed under the tap?

Q2 Which of the two carbohydrates passed through the Visking tubing into the surrounding water?

Q3 If the wall of the intestine has similar properties to those of Visking tubing, what must happen to the starch before the body can use it?

Q4 What other food substances do you think are likely to need similar treatment to the starch?

Q5 What food substances might pass through the gut wall without any such treatment?

9.2 Making big molecules smaller

In the laboratory, big molecules, such as starch, can be broken down into smaller molecules in a number of ways. One of these involves the use of dilute acid and heat.

1 Put 5 cm³ of 1 per cent starch suspension in a test-tube and add 10 drops of dilute hydrochloric acid.
2 In a second test-tube put another 5 cm³ of starch suspension and add 10 drops of distilled water.
3 Place both test-tubes in a water bath of boiling water.
4 At intervals of 5 minutes, test samples from each test-tube with iodine solution.
5 When the iodine solution no longer changes colour, add a little sodium bicarbonate slowly until the fizzing stops and then carry out a Benedict's test.

Q1 Can you suggest what has happened to the starch molecules?

To get chemical reactions to proceed quickly, a chemist sometimes breaks up solids by grinding them in a mortar. He also mixes solids and liquids thoroughly together. You do both these things with most of your food. You chew it well before swallowing it. You also mix it with a watery fluid called *saliva*, which is produced by the *salivary glands* which open into the mouth.

To investigate the action of saliva on starch, carry out the following experiment. Read *all* the instructions carefully before you start. *Figure 125* may also help you.

Figure 125
An experiment to investigate the
action of saliva on starch.

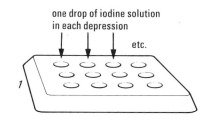

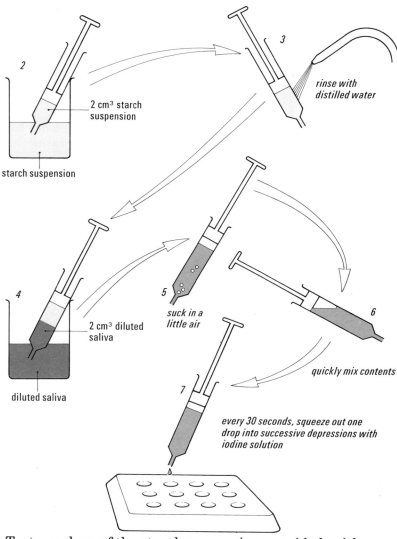

1 Test one drop of the starch suspension provided, with
iodine solution, on a white dish or spotting tile.
2 Rinse out your mouth with about 20 cm³ of distilled water;
spit this into the sink.
3 Rinse out your mouth a second time with 20 cm³ of distilled
water. This time keep the water in your mouth for half a
minute, moving it around with your tongue. Collect this
diluted saliva in a beaker.

4 Place *one* drop of iodine solution in each of the depressions on a spotting tile.

5 Use a clean 5 cm^3 syringe to take up 2 cm^3 of the starch suspension. Rinse the outside of the syringe with distilled water.

6 Now take up 2 cm^3 of dilute saliva into the same syringe. Draw in a little air and quickly and thoroughly mix the contents of the syringe by inverting it two or three times.

7 Hold the syringe vertically and *immediately* squeeze the plunger *carefully* to expel one drop of the mixture into the first drop of iodine solution on the tile. Do not let the tip of the syringe touch the iodine solution. Note the time.

8 Continue to test single drops of the mixture at intervals of 30 seconds until the iodine solution shows no further change in colour.

Q2 What appears to have happened to the starch during the course of this experiment?

Q3 What do you think was the purpose of rinsing out your mouth the first time?

9 Now expel 1 cm^3 of the remaining mixture in the syringe into a test-tube. Test this for the presence of reducing sugars, using Benedict's solution in the usual way.

10 Ten minutes later repeat step 9.

Q4 What conclusions can you draw from the results of the tests in steps *9* and *10*?

Q5 What test should you perform on the original starch suspension before you can be confident of your answer to question 4?

11 Place about 5 cm^3 of the remaining dilute saliva in a test-tube and boil it gently for about 10 seconds. Leave it to cool.

12 Repeat steps *4* to *7*, but this time mix boiled saliva with the starch suspension in a clean syringe. Continue to test single drops of this mixture for as long as you did with the unboiled saliva and starch.

Q6 How are the properties of saliva affected by boiling?

Q7 Saliva normally acts on starch at body temperature (37°C). If you had carried out the whole experiment at this temperature, what difference would you expect this to have made to the results you obtained?

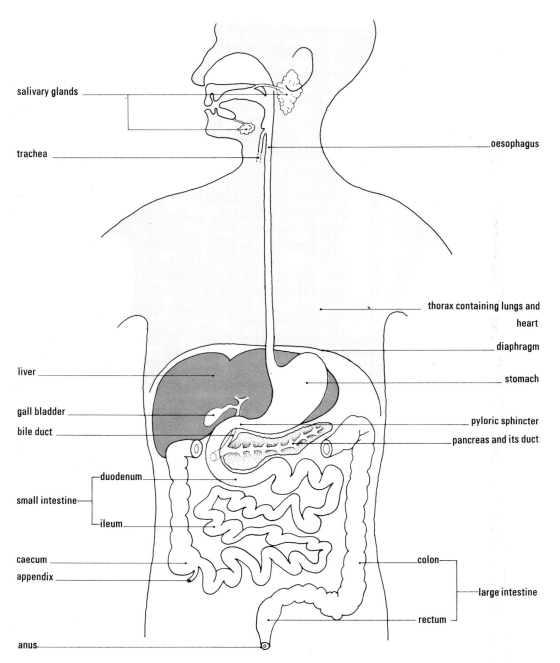

salivary glands

trachea

oesophagus

thorax containing lungs and heart

diaphragm

liver

stomach

gall bladder

pyloric sphincter

bile duct

pancreas and its duct

duodenum

small intestine

ileum

caecum

colon

appendix

large intestine

rectum

anus

9.21 Biological catalysts

Chemists often use *catalysts* to speed up reactions and let them take place more rapidly at lower temperatures than would otherwise be required. Catalysts also exist in every living organism; they are called *enzymes*. Nowadays enzymes are given names which show where they come from and what they do. The main constituent of starch is called amylose and an enzyme which acts on it is called an *amylase* – the ending, *-ase*, is used for all enzymes. So the enzyme in saliva is called *salivary amylase*. Each enzyme

Figure 126 *(opposite)*
A diagram of the human digestive system. The horizontal part of the large intestine has been cut away and the small intestine considerably shortened.

can catalyse only one chemical reaction. Saliva has no digestive effect on anything but starch. In the process of digestion, a succession of digestive enzymes get mixed with the food so that all the large molecules get broken down to small ones. For example, the stomach produces a *gastric juice* which contains the enzyme *gastric protease* (formerly called pepsin). This enzyme starts the process of protein digestion which is completed by other enzymes in juices acting in the intestines; the end product in this case is a mixture of amino acids. Digestion is only one process in which enzymes operate. All the activities in your body involve chemical changes which are catalysed by enzymes – in fact, every living cell contains large numbers of different enzymes. When you cut yourself, the blood soon clots; this is brought about by enzymes. When food goes bad, the changes in it are made by enzymes *produced* by microbes.

9.3 The human digestive system

The alimentary canal or gut is a tube running from the mouth to the *anus*. In an adult it is about ten metres long, most of its length being accounted for by the *small* and *large intestines* (*figure 126*). As food passes along the gut, digestive glands secrete juices containing the enzymes which catalyse the chemical changes involved in digestion.

9.31 How food is moved along the gut

In the wall of the gut are two layers of muscle fibres. One set is arranged along the length of the gut and the other in a circular manner (see *figure 127*). These muscles work in

Figure 127
A diagram to show the two layers of muscle fibres in the wall of the gut.

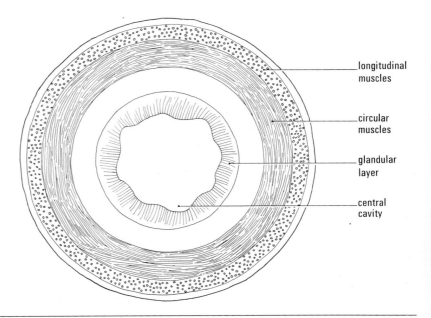

longitudinal muscles

circular muscles

glandular layer

central cavity

conjunction with each other to make the gut shorter and fatter or longer and thinner.

Q1 Which set of muscles do you think must contract to make the gut *a* shorter and *b* thinner?

The circular muscles contract in succession, so pushing the food along. This process is called *peristalsis*. This is what happens in the oesophagus as soon as you have swallowed some food (see *figure 128*).

Figure 128
Peristalsis: how food is moved along the gut.

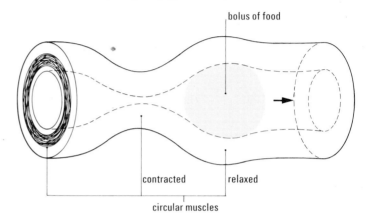

bolus of food

contracted   relaxed

circular muscles

Muscular movements also serve to mix the food with the digestive juices. Peristalsis can sometimes be observed in the gut of a mammal dissected immediately after death.

At certain points along the gut there are special thickenings of the circular muscles to form valves called *sphincters*. One of these is at the point where the stomach leads to the *duodenum* and is called the *pyloric* sphincter. It is normally relaxed and open, but it closes each time a

Figure 129
X-ray pictures of a human gut (front view), taken after the subject had swallowed a suspension of barium meal.
a Ten minutes later.
b Eighteen hours later.
Photographs, Ilford Ltd.

a   b

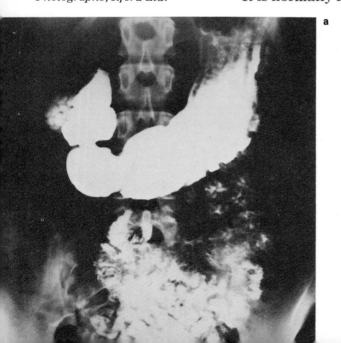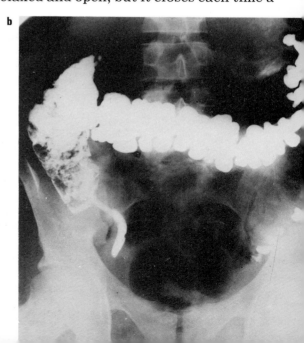

wave of contraction passes along the stomach. In this way it prevents large lumps of food from leaving the stomach, while letting semi-liquid material flow into the duodenum.

The movement of the human gut can be studied through X-rays. If someone eats a meal containing barium sulphate (an insoluble substance which is opaque to X-rays) the part of the gut occupied by the food can be studied on the X-ray screen or photographed (see *figure 129*). In this way the movement of materials along the gut can be timed.
A typical meal shows results like those below:
a Mouth to stomach: liquids – a few seconds; dry solids – 1–2 minutes.
b Stomach empties in approximately 6 hours.
c Small intestine empties in approximately 8–9 hours.

Q2   Which parts of the gut show up in *figure 129*?

9.32 Studying digestion

It is easy to test salivary digestion, but how can one examine fluids from the stomach or intestines? The French scientist Réaumur (1683–1757) fed a kite on pieces of sponge which the bird later vomited up (kites vomit any indigestible material). Using other kinds of bird, Réaumur fed them sponges attached to long threads so that they could be withdrawn later. Squeezing out the gastric juice, he found that it softened meat.

A more direct method is to pass into the stomach a length of flexible tubing, so that a syringe connected to the upper end can be used to withdraw a sample of the gut contents. This device (a stomach pump) is also used to recover poisons after they have been swallowed. Sometimes it is possible to study the digestive process when a part of the gut is exposed by surgery or as a result of an accident. You will find an account of the most famous example of such a study in the Background reading at the end of the chapter.

More recently the conditions within the human stomach have been detected and relayed back to an observer by minute, pill-like radio transmitters which subjects can swallow. In addition, work with experimental animals has helped in unravelling the complicated details of digestion.

9.33 The digestive process

Chewing, and muscular movement of the gut, break up large pieces of solid food and mix the fragments with liquid food and digestive juices. In this semi-liquid condition the various enzymes of the different digestive juices catalyse the chemical breakdown of starch to sugars and of protein to amino acids. Both these 'end-products of digestion' can be absorbed.

Fats present a peculiar problem, for they do not mix readily with water. Churning in the stomach produces a coarse emulsion – large fat droplets suspended in water. When the fat passes out into the duodenum it becomes mixed with *bile* (a juice produced in the liver and stored temporarily in the gall bladder). The bile is the chief emulsifying agent in breaking up the fat into minute droplets. If these are of a sufficiently small size (0.2 mm), they can pass unchanged through the gut wall. Some fat, however, is digested by enzymes to produce simpler compounds called fatty acids and glycerol.

9.34 Absorbing the useful material

You may have seen in Chapter 5 that the large surface area of the lungs allows efficient and rapid absorption of oxygen. For the same reasons the lining of the gut needs a large surface area for efficient and rapid absorption of the soluble products of digestion. The length of the intestine helps in providing a large surface area, but this is greatly increased by the folds of the lining and by the minute projections called *villi* (singular – villus) which cover these folds (see *figures 130, 131,* and *132*). Each villus is only about 1 mm long and less than 0.25 mm thick. There are between 20 and 40 on each mm^2 of intestinal lining. Thus, the total area exposed to the gut contents is very large indeed and has been estimated to be 300 m^2. The villi also contain muscle fibres which enable them to be moved.

Figure 130 *(below, left)*
How the surface of the gut wall is increased.

Figure 131 *(below, right)*
Part of the wall of the human intestine, showing villi. It has been treated chemically to make the villi easier to see.
Photograph, Professor C. C. Booth, Royal Postgraduate Medical School, Hammersmith Hospital.

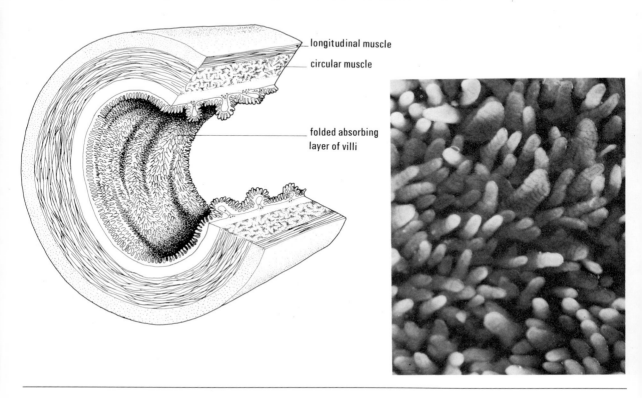

longitudinal muscle

circular muscle

folded absorbing layer of villi

Living things in action

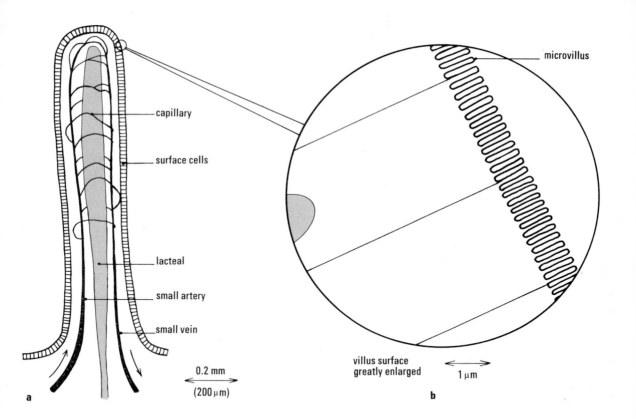

capillary

surface cells

lacteal

small artery

small vein

microvillus

villus surface
greatly enlarged

0.2 mm

(200 µm)

1 µm

a

b

Figure 132
a The structure of a single villus.
b The surface layer of cells,
greatly enlarged.

Q3 Estimate the area of the floor of the room you are in. Is it more or less than 300 m²?

Electron micrographs show that each of the cells forming the surface of a single villus has a 'brush border' of about 1000 projections, each 1 µm long, called microvilli (see *figure 132*). These increase the surface area even further.

Each villus contains a network of small blood vessels and a central tube called a *lacteal* (see *figure 132*). Most digested food diffuses through the walls of the villi into the blood stream, but fat droplets pass into the lacteals, which unite to form lymphatic vessels. These eventually discharge their contents into the blood stream.

All the blood from the gut passes into one large vessel – the *hepatic portal vein* – which leads to the liver. Here much of the food is stored and processed before being passed on to the rest of the body for use. For instance, excess glucose is converted to glycogen, which is stored in the liver. The glycogen is converted back to glucose again when needed to replace that which has been used by the respiring cells of the body. Thus the level of glucose in the circulating blood is always kept fairly constant.

9.35 Eliminating the residue

Digestion and absorption are very efficient processes, so by the time the food material reaches the end of the ileum there is hardly any digested food left unabsorbed and much of the water has been absorbed also. However, the cellulose and fibrous materials of plant food are largely unchanged, since none of the digestive enzymes has any effect on these. Bacteria, which are abundant in the guts of herbivores, can break these substances down. The herbivores are then able to absorb the products of this digestion. Humans, however, do not have the appropriate bacteria and so do not gain any advantage from this. The main function of the large intestine is to conserve water by absorbing as much of it as possible, so that the eliminated material (*faeces*) is often fairly dry. Faeces consist largely of dead and living bacteria, together with undigested solid matter and dead cells worn away from the gut lining. This material is eventually expelled from the body through the anal sphincter by muscular contractions of the rectum – this is *defecation*. Indigestible material (mainly cellulose) is referred to as *roughage* and most nutritionists accept that a certain amount of it in the diet aids defecation, *i.e.* prevents constipation.

A lucky escape

Important scientific discoveries sometimes result from a chance event, if a competent observer studies it.
A major breakthrough in the study of digestion resulted from a near fatal accident in the American backwoods. It happened in Mackinac, near the junction of Lake Michigan and Lake Huron. This small settlement had an army fort and a store of the American Fur Company, where trappers came in to do business. On 6 June 1822, some people were standing in the store when one of them accidentally discharged a shotgun, wounding Alexis St Martin, a young French Canadian who was standing about a metre away. An eye witness takes up the story:

'Doctor Beaumont, the surgeon of the fort, was immediately sent for and reached the wounded man in a very short time, probably three minutes. We had just gotten him on to a cot and were taking off some of his clothing. After the doctor had extracted part of the shot, together with pieces of clothing, and dressed his wound carefully, he left him, remarking: "The man cannot live thirty-six hours; I will come and see him by and by." In two or three hours he visited him again, expressing surprise and finding him doing better than he had anticipated. The next day, after getting out more shot and clothing and cutting off ragged edges of the wound he said he thought he would recover.'

Figure 133
William Beaumont.
Photograph, The Mansell Collection.

Living things in action

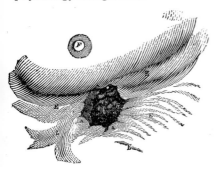

Figure 134
Drawing of the hole in Alexis
St Martin's stomach and side.
From Beaumont, W. (1833)
Experiments and observations on
the gastric juice and the
physiology of digestion.

This engraving represents the appearance of the
aperture with the valve depressed.

William Beaumont was the army doctor who treated
St Martin. In a 'memorial' later presented to the Senate
and House of Representatives, he wrote:
'The wound was received just under the left breast, and
supposed at the time to have been mortal. A large
portion of his side was blown off, the ribs fractured and
openings made into the cavities of the chest and abdomen, through which protruded portions of the lungs and
stomach, much lacerated and burnt, exhibiting altogether an appalling and hopeless case. The diaphragm was lacerated and a perforation made directly
into the cavity of the stomach through which food was
escaping at the time your memorialist was called to his
relief. His life was at first wholly despaired of, but he
very unexpectedly survived the immediate effects of the
wound, and necessarily continued a long time under the
constant professional care and treatment of your
memorialist, and, by the blessing of God, finally recovered his health and strength.'

It took a year before St Martin was fully recovered
and by a strange chance, when it healed the edges of the
hole in the stomach became attached to the edges of the
external wound. Beaumont tried several times to close the
hole (a *fistula*), but it was too big (see *figure 134*).

By now St Martin had been declared a 'common pauper' by
the local authorities, who resolved to transport him back
to Canada. Beaumont realized that the journey might well
be too much for St Martin, and took him into his service.
In 1825 he started to experiment on the contents of St
Martin's stomach. The following extract is taken from
Beaumont's own account:

'August 7. At 11 o'clock a.m., after having kept the lad
fasting, for seventeen hours, I introduced the glass tube
of a thermometer (Fahrenheit's) through the perforation,
into the stomach, nearly the whole length of the stem, to
ascertain the natural warmth of the stomach. In fifteen
minutes, or less, the mercury rose to 100° [38°C] and
there remained stationary. . . .

'I now introduced a . . . tube, and drew off one ounce
[28 g] of pure gastric liquor, unmixed with any other
matter, except a small proportion of mucus, into a three
ounce vial [85 g]. I then took a solid piece of boiled,
recently salted beef, weighing three drachms [10 g],
and put it into the liquor in the vial; corked the vial
tight, and placed it in a saucepan, filled with water,
raised to the temperature of 100° and kept at that point
on a nicely regulated sand bath. In forty minutes
digestion had distinctly commenced over the surface of
the meat. In fifty minutes the fluid had become quite

opaque and cloudy; the external texture began to separate and become loose. In sixty minutes, chyme began to form.

'At 1 o'clock, p.m. (digestion having progressed with the same regularity as in the last half hour) the cellular texture seemed to be entirely destroyed, leaving the muscular fibres loose and unconnected, floating about in fine small shreds, very tender and soft.

'At 3 o'clock, the muscular fibres had diminished one half, since last examination at 1 o'clock.

'At 5 o'clock, they were nearly all digested; a few fibres only remaining.

'At 7 o'clock, the muscular texture was completely broken down; and only a few of the small fibres floating in the fluid.

'At 9 o'clock, every part of the meat was completely digested.

'The gastric juice, when taken from the stomach, was as clear and transparent as water. The mixture in the vial was now about the colour of whey. After standing at rest a few minutes, a fine sediment, of the colour of the meat, subsided to the bottom of the vial.

'. . . At the same time that I commenced the foregoing experiment, I suspended a piece of beef, exactly similar to that in the vial . . . into the stomach, through the aperture.

'At 12 o'clock, m., withdrew it, and found it about as much affected by digestion as that in the vial; there was little or no difference in their appearance. Returned it again.

'At 1 o'clock p.m. I drew out the string; but the meat was all completely digested, and gone.'

Q1   Notice that Beaumont kept both samples of gastric juice at the same temperature. Suggest a reason why the meat was digested more quickly in the stomach than in the vessel (vial).

During this year the army transferred Beaumont to Plattsburgh, New York. He took St Martin along with him, hoping to exhibit him to other doctors there. But St Martin seems to have had enough of being a medical specimen, for he slipped off without telling his master. Four years went by before Beaumont finally learnt of St Martin's whereabouts – he was now married with two children, and working in Canada for the Hudson Bay Fur Company.

Beaumont himself was now stationed at Fort Crawford, Upper Mississippi, and he arranged to transfer St Martin and his family to the same area, where St Martin was given a job with the American Fur Company. His fistula had hardly changed in the four years, and Beaumont continued his experiments, in return for which St Martin received free board and lodgings plus 150 dollars a year. Between 1819 and 1833, when St Martin took his family back to Canada for good, Beaumont carried out dozens of experiments, published in his book *Experiments and observations on the gastric juice and the physiology of digestion*. You can gain some idea of Beaumont's detailed approach in *table 18*, a short extract from his results.

Articles of diet	Time of chymification*			
	in stomach		in vials	
	Hrs.	min.	Hrs.	min.
eggs, hard boiled	3	30	8	00
eggs, soft boiled	3	00	6	30
eggs, fried	3	30		
eggs, roasted	2	15		
eggs, raw	2	00	4	15
eggs, whipped	1	30	4	00
salmon, salted, boiled	4	00	7	45
oysters, raw	2	55	7	30
oysters, roasted	3	15		
oysters, stewed	3	30	8	25
bone, beef, solid			80	00
sponge cake	2	30	6	15

* 'Chymification' is the digestion of solid food to *chyme*, a soup-like liquid.

Table 18

Beaumont's work had a wide influence in America and Europe, and much of it remains in present-day textbooks. It was said of Beaumont:

'It would be difficult to point out any observer who excels him in devotion to truth, and freedom from the trammels of theory or prejudice. He tells plainly what he saw and leaves every one to draw his own inferences, or where he lays down conclusions he does so with a degree of modesty and fairness of which few perhaps in his circumstances would have been capable.'

Plants and the atmosphere

10.1 Plants and carbon dioxide

Investigations in Chapter 4 show that all living organisms exchange gases with their environment as a result of respiration. Even if you have not studied that chapter you will probably know that human beings and other animals give out carbon dioxide to the atmosphere around them. The respiration of living organisms is responsible for adding vast quantities of carbon dioxide to the Earth's atmosphere; one estimate is that $500\,000\,000\,000\,(= 5 \times 10^{11})$ tonnes of this gas are produced each year.

Q1 Following on from this, what would you expect to happen to the concentration of carbon dioxide in the air over a period of days or weeks?

However, when samples of air are analysed, provided they are collected away from industrial areas and well clear of the ground, they are found to contain, on average, 0.03 per cent of carbon dioxide.

Q2 What conclusion follows from this fact?

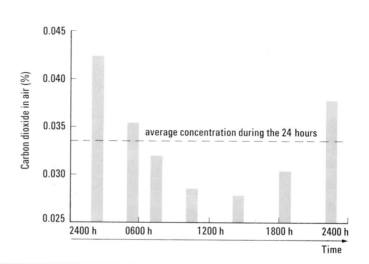

Figure 135
Variation in the concentration of carbon dioxide in air taken from between the leaves in a field of long grass. The crop was about 85 cm tall, and the air was pumped from a height of 40 cm above the soil.
After Monteith, J. L. (1962)
Neth. J. agric. Sci. **10,** *5 334–346.*

Living things in action

Any answer to question 2 is best regarded as a tentative guess (a *hypothesis*) if you have no experimental data to support it. In one relevant experiment, the results of which are shown in *figure 135*, a scientist accurately measured the carbon dioxide concentration of the air in the middle of some long grass. Notice that he took samples at intervals during a single day. Note the time of day when the amounts of carbon dioxide increased and decreased. The change in the quantity of carbon dioxide must be because of some process taking place.

Q3 Examine this list of factors and suggest which ones could be associated with the change in carbon dioxide concentration: *a* activity of animals *b* temperature *c* rainfall *d* sunlight *e* wind *f* industrial pollution.

Plants and animals in a closed community
You can now investigate the exchange of gases in a closed community of animals and plants. The following experiment uses aquatic animals and plants.
Eight tubes are set up as shown in *figure 136*.
Tubes A and E: animal material alone in distilled water.
Tubes B and F: plant material alone in distilled water.
Tubes C and G: plant and animal material together in distilled water.
Tubes D and H: distilled water alone.
Tubes A to D are kept in a well lit place and
Tubes E to H are kept in the dark for at least 12 hours.

Figure 136
An experiment to show how animals and plants in a closed community affect the environment around them.

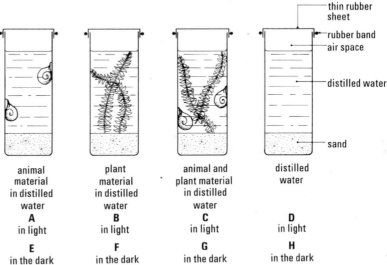

animal material in distilled water	plant material in distilled water	animal and plant material in distilled water	distilled water
A in light	**B** in light	**C** in light	**D** in light
E in the dark	**F** in the dark	**G** in the dark	**H** in the dark

1 After this time, examine the tubes and their contents and record all the observations you can.
2 Use an indicator to detect any changes in carbon dioxide concentration.
Before you use the indicator, first compare the colour

changes of the indicator in tubes which will show the colour of the indicator with

a normal air

b normal air $+ CO_2$

c normal air $- CO_2$

Now complete the investigation with these closed communities by injecting 2 cm³ of indicator solution through the rubber cap of each tube. Record your observations and deductions in a table like the one shown in *table 19*.

Tube	Observation	Colour change of indicator	Deductions
A			
B			
etc.			

Table 19

Q4 What can you conclude about the exchange of carbon dioxide between animals and plants in the light?

Q5 Since both plants and animals produce carbon dioxide at night, what must you conclude about its uptake by plants during the day?

10.11 Testing a hypothesis

The experimental results shown in *figure 135* and your own results may have led you to guess something like this: 'Green plants take up carbon dioxide from the atmosphere but only when they are exposed to light.' A hypothesis like this remains a guess until it has been tested. You can test it by using an indicator solution again.

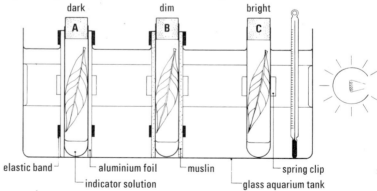

Figure 137
How green leaves affect indicator solution at different light intensities.

This is what you do:

1 Wash out three corked test-tubes with distilled water and then with a little indicator solution. Take care not to breathe near the tubes and do not put your thumb over the end of the open tube – the acids in the sweat of your thumb will affect the indicator.

2 Using a similarly washed syringe or pipette, quickly run 2 cm³ of indicator solution to the bottom of each tube and replace the bung.

3 Label the tubes A, B, and C. Take all the tubes to the leaves you are to use, and with forceps slip one leaf quickly inside each tube, replacing the bung as before. (See *figure 137*.)

4 Arrange muslin or aluminium foil around the appropriate tubes and then support them in a water bath kept at a constant temperature, in a well lit position.

5 One group should also set up tubes with no leaves (labelled D).

6 Record the time at which the tubes were set up, and then give each tube a shake every five minutes so that the indicator solution makes better contact with the air above it. Do not get any indicator on the leaves. Note any colour changes by comparing the tubes containing leaves with a tube with no leaf.

7 The last time you look at the tubes, remove the leaves and look through the indicator solution at a well lit white surface. Compare the final colour with the control tubes.

You should obtain clear results in an hour, but the time taken will depend on the brightness of the light. Full sunlight is best but, of course, is not always available. If you use electric light bulbs they give out heat as well as light – but sunlight also has a heating effect. But the heating effects are reduced, either by standing the tubes in a water bath or by placing a heat filter between the tubes and the light source.

A similar experiment can be performed using water plants instead of leaves. Wash the plants well before you use them and use just enough indicator solution to cover them. You can assume that the indicator solution does not affect the water plants during the experiment. Again, set up other tubes with no water plants.

Figure 138
How water plants affect indicator solution at different light intensities.

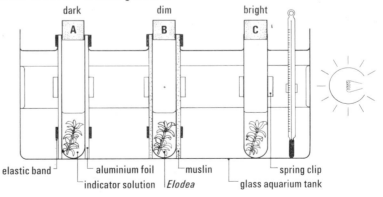

Q6 Why should the tubes only contain a small volume of indicator solution?

Q7 What is the point of setting up tubes without leaves or water plants in them?

Compare your results with those of the rest of the class.

Q9 Explain how the results contradict or agree with the hypothesis suggested at the beginning of this section.

10.12 A closer look at gas exchange in a green plant

All living cells require energy, which they obtain by breaking down food materials to simpler ones; this process, respiration, is studied in Chapter 6. A living plant requires energy during the day and night so, if the temperature does not change greatly, the cells in the plant's leaves and other organs will produce carbon dioxide at the same rate

Figure 139

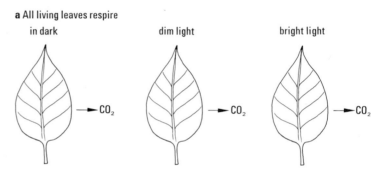

a All living leaves respire

in dark   dim light   bright light

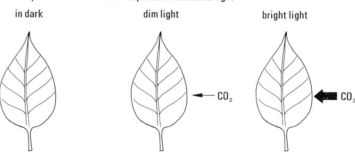

b Absorption of carbon dioxide depends on available light

in dark   dim light   bright light

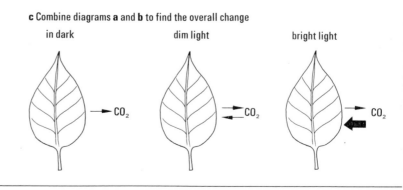

c Combine diagrams **a** and **b** to find the overall change

in dark   dim light   bright light

whether in the dark, in dim light, or in bright light. *Figure 139* represents the carbon dioxide balance of the leaf and will help to explain the results of the previous experiments to you.

10.13 Tracing the path of carbon atoms

Much recent research on the uptake of carbon atoms involves the use of radioactive atoms. Most carbon atoms have a mass of 12, but a rarer form exists with a mass of 14. These are called isotopes of carbon. The ^{12}C isotope is stable but the ^{14}C isotope is radioactive. A plant is supplied with carbon dioxide, the carbon atoms of which are of the ^{14}C isotope. A counter is later used to find whether the plant has taken up more carbon dioxide in bright light than

Figure 140
The effect of light and darkness on the uptake of labelled carbon dioxide ($^{14}CO_2$) by *Chlorella*.

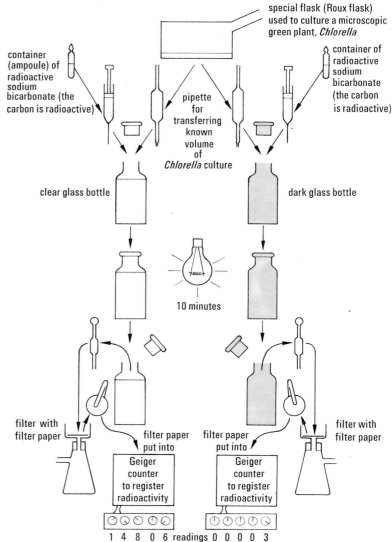

Figure 141
Cells of *Chlorella* (× 1000).
Photograph, Philip Harris Biological Ltd.

it did in the dark. The procedure is illustrated in *figure 140* and you may be able to watch a film loop which shows the experiment being carried out. A microscopic green plant, the alga *Chlorella*, is used in the experiment (see *figure 141*).

The following questions may be answered from the film loop or from *figure 140*.

Q10   At the end of the experiment, what is the reading on each Geiger counter and what do these results suggest?

Q12   This experiment tests the same basic hypothesis as your experiment with indicator solution in section 10.11. Does it give the same answer, or a different one?
In what ways does the use of ^{14}C give more information?

10.2 Plants and oxygen

You now have evidence that it is the interaction of plants and animals which keeps the concentration of carbon dioxide in the atmosphere at a constant level.

We must also try to explain the relative constancy of the oxygen content of the Earth's atmosphere at about 21 per cent by volume. The work in Chapter 4 provides the evidence for the fact that oxygen is removed when living organisms respire. We now have to investigate the means by which the oxygen is put back into the atmosphere.

It would be quite possible to follow changes in the oxygen content of the air in a field of grass, just as the carbon dioxide concentration was followed, as shown in *figure 135*. Instead we have chosen to examine what happens to water, before you investigate the atmosphere.

Study *figure 142* which shows certain changes in the water of a river during twenty-four hours.

Look at the graphs which show changes in the pH (acidity) and oxygen concentration in the water.

Q1   What is the most likely cause of these changes and how does this graph relate to the one in *figure 135*?

Now look at the graph of the amount of oxygen dissolved in the river water and note the times of day or night when the concentration is decreasing and the times when it is increasing.

Q2   How can you explain these changes in the oxygen content of the water?

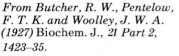

Figure 142
Some changes in the water of the
River Lark during twenty-four
hours.
From Butcher, R. W., Pentelow,
F. T. K. and Woolley, J. W. A.
(1927) Biochem. J., *21 Part 2,*
1423–35.

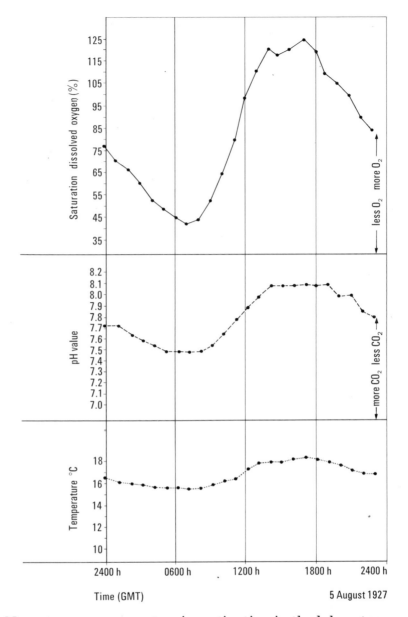

Now you can carry out an investigation in the laboratory.
1   Take a beaker of pond water containing some sprigs of a
    water plant (*e.g. Elodea*).
2   Put it in a shaded part of the laboratory and then switch on
    a bench lamp which is about 50 cm away from the beaker.
3   Watch to see what happens and devise a way of measuring
    the process.
4   Now bring the lamp closer, to 25 cm away. Again watch
    the beaker and record any change that may happen.

Q3   What exactly can you conclude about light intensity and
     the production of gas, from the results of this experiment?

10.21 Analysis of the gas

To be really sure what gas the water plant produced you must collect some and analyse it. It would be interesting to find out, for instance, if its composition is similar to that of the atmosphere.

Figure 143 shows how a fairly large sample of the gas can be collected; this apparatus needs to be set up a few days before you are going to carry out the analysis.

Figure 143
How to collect gas from water plants for analysis.

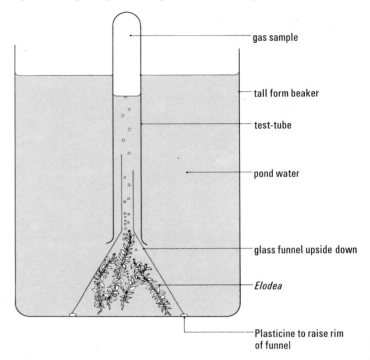

gas sample

tall form beaker

test-tube

pond water

glass funnel upside down

Elodea

Plasticine to raise rim of funnel

To analyse the gas you can use one of the techniques suggested in section 4.2, although small volumes of the gas are probably best analysed with a capillary tube rather than a gas syringe (see *figure 52,* page 58). If you have not carried out the investigation in 4.2 you will need to study that section carefully, and even if you have it would be best to revise the main points.

When you have carried out the gas analysis, examine your results and compare them with those obtained by the rest of the class. You will have to think rather carefully about the figures you obtain.

Q4 Did you detect any change with potassium hydroxide solution? What does this mean?

Living things in action

Q5 What percentage of oxygen was present and how does it compare with the percentage normally in atmospheric air?

Q6 What control should be carried out for this experiment?

Q7 What can you now suggest is the means whereby oxygen is put back into the atmosphere?

Background reading

Some historical evidence

The special role of green plants in nature has attracted the interest of scientists for a long time. Here are three accounts taken from scientific papers published in the seventeenth and eighteenth centuries. They will give you an idea of how much was known about this subject when they were written.

By comparing the results of the three scientists' work with your own results you will be able to see whether or not you have arrived at the same conclusions and whether there are other aspects to be investigated.

'By experiment, that all vegetable matter is totally and materially of water alone'
by Jean-Baptiste van Helmont (1577–1644). From *Ortus medicinae,* the collected edition of the author's works, first published in Amsterdam in 1668.*

'That all vegetable [matter] immediately and materially arises from the element of water alone I learned from this experiment. I took an earthenware pot, placed in it 200 lb [about 100 kg] of earth dried in an oven, soaked

Figure 144
Van Helmont's experiment, described by him in *Ortus medicinae.*

5 lb shoot + 200 lb dry soil + 5 years with only = 169 lb 3 oz tree + 199 lb 14 oz dry
                                    a supply of rain   soil
                                    water for growth

*(Excerpt taken from *Great experiments in biology,* edited by Gabriel, M. L., and Fogel, S.; translated by Professor N. Lewis.)

this with water, and planted in it a willow shoot weighing 5 lb [about 2 kg]. After five years had passed, the tree grown therefrom weighed 169 lb and about 3 oz [about 77 kg]. But the earthenware pot was constantly wet only with rain or (when necessary) distilled water; and it was ample [in size] and embedded in the ground; and, to prevent dust flying around from mixing with the earth, the rim of the pot was kept covered with an iron plate coated with tin and pierced with many holes. I did not compute the weight of the deciduous leaves of the four autumns. Finally, I again dried the earth of the pot, and it was found to be the same 200 lb minus about 2 oz. Therefore, 164 lb [74 kg] of wood, bark, and root had arisen from the water alone.'

Clearly the willow tree increased in mass considerably. Van Helmont considered that this came from the water.

Q1   Do you agree or have you any other explanation?

'Observations on different kinds of air'
by Joseph Priestley. Taken from *Philosophical Transactions* of the Royal Society (1772).

'I flatter myself that I have accidentally hit upon a method of restoring air which has been injured by the burning of candles, and that I have discovered at least one of the restoratives which nature employs for this purpose. It is vegetation. In what manner this process in nature operates, to produce so remarkable an effect, I do not pretend to have discovered; but a number of facts declare in favour of this hypothesis. I shall introduce my account of them, by reciting some of the observations which I made on the growing of plants in confined air, which led to this discovery.

'One might have imagined that, since common air is necessary to vegetable, as well as to animal life, both plants and animals had affected it in the same manner, and I own I had that expectation, when I first put a sprig of mint into a glass-jar, standing inverted in a vessel of water; but when it had continued growing there for some months, I found that the air would neither extinguish a candle, nor was it at all inconvenient to a mouse, which I put into it . . .

'Finding that candles burn very well in air in which plants had grown a long time, and having had some reason to think, that there was something attending vegetation, which restored air that had been injured by respiration, I thought it was possible that the same process might also restore the air that had been injured by the burning of candles.

'Accordingly, on the 17th of August 1771, I put a sprig of mint into a quantity of air, in which a wax candle had burned out, and found that, on the 27th of the same month, another candle burned perfectly well in it. This experiment I repeated, without the least variation in the event, not less than eight or ten times in the remainder of the summer. Several times I divided the quantity of air in which the candle had burned out, into two parts, and putting the plant into one of them, left the other in the same exposure, contained, also, in a glass vessel immersed in water, but without any plant; and never failed to find, that a candle would burn in the former, but not in the latter. I generally found that five or six days were sufficient to restore this air, when the plant was in its vigour.'

Q2   Which of the experiments you have carried out support Priestley's findings?

Q3   Priestley stated that 'restored' air was not at all 'inconvenient' to a mouse which he put in it. Can you explain what he was talking about in terms of the results of the experiments you have been studying?

'Experiments upon vegetables'
by John Ingen-Housz. Extracts from *Experiments upon vegetables, discovering their great power of purifying the common air in the sun-shine, and of injuring it in the shade and at night* (1779).

'I was not long engaged in this enquiry before I saw a most important scene opened to my view: I observed, that plants not only have a faculty to correct bad air in six or ten days, by growing in it, as the experiments of Dr Priestley indicate, but that they perform this important office in a compleat manner in a few hours; that this wonderful operation is by no means owing to the vegetation of the plant, but to the influence of the light of the sun upon the plant . . . that this office is not performed by the whole plant, but only by the leaves and the green stalks that support them . . . Two handfulls of leaves of French beans were put in a jar of a gallon; it was kept inverted upon a dish, and some water poured upon it; next morning I found the air so much fouled that a candle could not burn in it . . . After this I replaced it again in the sun till five in the afternoon, when I found the air so much mended as to be equal in goodness to common air . . . I put some green stalks of a willow-tree, the leaves being stripped off, in a gallon jar filled with pump-water; the jar was exposed, inverted, as ordinary, upon a wall in a warm sun-shine during four hours. They became most beautifully covered with an infinite number of round air-bubbles . . .'

This was the stage the investigations had reached by the end of the eighteenth century.

Q4   How did the results of the experiments done by Ingen-Housz add to the information Priestley obtained?

Q5   Which of the problems, if any, which you have investigated, give similar findings to those of Ingen-Housz?

11

Plants, food, and light energy

11.1 Green plants and food webs In Chapter 8, sheep and locusts are considered as examples of herbivores (plant-eating animals). Herbivores themselves are preyed upon by carnivores (flesh-eating animals). As man eats both plant and animal food he is called an *omnivore*. In any community we can arrange the living organisms in a series of 'eaters' and 'eaten' to form a *food chain*. A simple example is

grass → beef cattle → man

Since the grass seems to produce the food materials that the animals eat (consume) the terms *producer* and *consumer* are also used and the general pattern for all food chains may be written:

producers → primary consumers → secondary consumers
(green plants) (herbivores) (carnivores)

But the pattern is usually more complicated than this. The single chain, for instance, can have more than two links:

cabbage → large white → thrush → sparrowhawk
　　　　　　caterpillar

Also, thrushes do not only eat large white caterpillars; they feed on a variety of other creatures including snails. Many plants are eaten by more than one species of herbivore and so the term 'food web' is used to describe the network of food chains.

Try working out food webs in a variety of habitats and then tracing the different food chains within them.

Q1 After considering several food chains what conclusion can you come to about the source of all food?

In van Helmont's experiment, described on page 157, he waited five years to measure the increased growth of the willow plant. If we are to carry out experiments on plant feeding we must find a more rapid way of deciding whether a plant is manufacturing food.

11.2 When do leaves contain food?

Of all the food tests in Chapter 7, the simplest is that for starch. If iodine solution is added to a piece of bread or potato, the starch present reacts to give a blue–black colour. However, in a material which is already strongly coloured, like a green leaf, the result can be seen better if the colouring is first removed.

Figure 145
How to test leaves for starch.

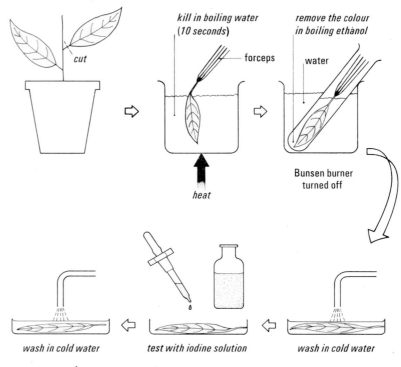

This is what you do (see *figure 145*):

1   Remove a leaf from a plant.
2   Hold the leaf with a pair of forceps and kill it by putting it in boiling water for at least 10 seconds.
3   Remove the green colouring by soaking the leaf in boiling ethanol. Use a beaker of hot water to heat the boiling-tube of ethanol. It **must not** be heated with a direct flame because ethanol is highly inflammable. *All* Bunsen burners must be turned out before ethanol is used.
4   When the leaf is colourless or pale yellow, wash it in cold water for a few seconds. The ethanol will be washed away, making the leaf less brittle.
5   Pour away the water and replace it with iodine solution. Any starch present will react with the iodine solution to give a blue–black colour while leaf tissue is stained yellow–brown. The leaf may then be washed again with water.

Now carry out this starch test on the following three leaves. Because the test involves removing all the green colour, it is a good idea to draw leaf C first so that you know which parts were green.

A – green leaf which has been growing in the light.
B – green leaf from a plant which has been kept in the dark for 48 hours.
C – variegated leaf which has been growing in the light.

Q1 What do the results tell you about the effect of light and darkness?

Q2 What do the results tell you about the importance of green colouring?

11.3 What happens to carbon atoms in a leaf?

Two key facts which have been established so far are:
1 Green plants in the light absorb carbon dioxide from their surroundings.
2 Green leaves which have been in the light often contain starch.

It looks as though carbon dioxide might be used to form starch. A simple way to represent this is:

$$\boxed{\text{carbon dioxide}} + \boxed{?} \rightarrow \boxed{\text{starch}}$$

If this is so then it should be possible to prevent starch formation by depriving leaves of carbon dioxide.

You can devise apparatus in which to grow green plants
a in the light and with carbon dioxide in the air around them;
b in the light but without any carbon dioxide;
c in darkness but with carbon dioxide.

To deprive the plants of carbon dioxide put an alkaline substance such as soda lime near the leaves in a closed container. Grow the plants under these conditions for a day or two and then test the leaves for starch. Also set up a control.

Q1 What deductions can you make from your results?

The question we might now ask is whether the change from carbon dioxide to starch is a simple process or whether there is a number of steps.

The chemical formula for carbon dioxide is CO_2. The structural formula for starch is shown in *figure 146*, which also shows the structural formula for the simple hexose sugar, glucose. After looking at these diagrams you may get a clue to a possible mechanism.

Figure 146
The structural formula for starch and glucose. (See also *figure 91*, page 105.)

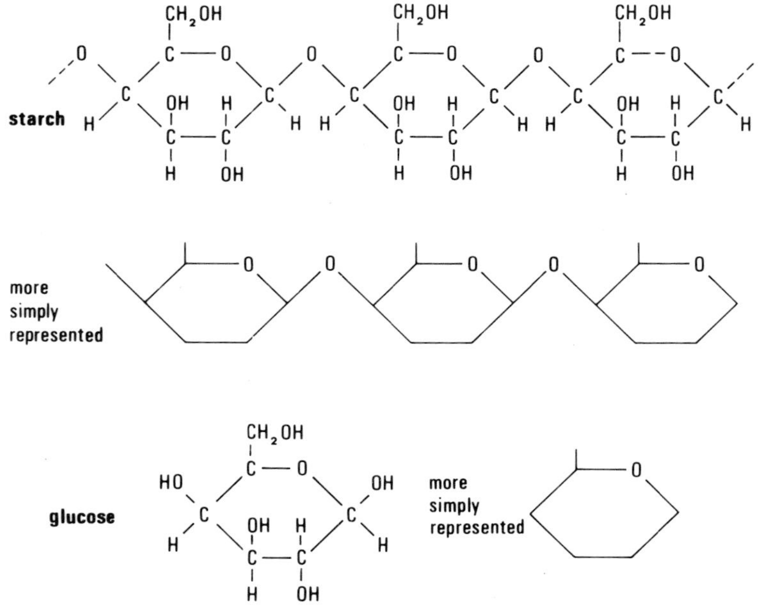

Q2   How would you complete the following diagram now?

| carbon dioxide | + | ? | → | | → | starch |

This raises another question: can starch be made from glucose? You can find the answer by carrying out an experiment. You will be provided with some Petri dishes containing agar to which some glucose-1-phosphate has been added. You are also given some 'juice' from a potato.

1   Test a drop of the potato 'juice' for the presence of starch.
2   Cut out four circles of agar with a cork borer.
3   Put the same amount of potato juice into each hole, but do not quite fill it.
4   After five minutes add some iodine solution to one hole and note the result.
5   Do this to the other holes in turn after 10, 15, and 20 minutes.
6   Carry out a suitable control experiment.

What answer do the results give to the question: can starch be made from glucose?

Suppose that you now tried to represent an overall picture of the change of carbon dioxide to glucose. You know that on the lefthand side you can write carbon dioxide and on the righthand side, glucose. In Chapter 10 there is evidence that oxygen is produced by green plants so that might also be written on the righthand side.

carbon dioxide → glucose + oxygen
CO_2 $C_6H_{12}O_6$ O_2

There are two problems: first, to account for the hydrogen which is added to the carbon atoms and, second, to discover the source of oxygen which escapes to the atmosphere.

11.4 A source of hydrogen and oxygen atoms

The answer to these two questions is remarkably simple. There is evidence that water, H_2O, is the source of both these kinds of atoms, being split up in some way so that its oxygen atoms escape to the atmosphere, while its hydrogen atoms are used by the plant to *reduce* the carbon dioxide. Adding hydrogen or removing oxygen is called *reduction*. Green plants therefore reduce atmospheric carbon dioxide to sugars.

Evidence about the source of oxygen has been obtained in the following way. Ordinary oxygen is composed of several isotopes (see page 153), the most abundant being ^{16}O. The heaviest isotope, ^{18}O, forms only a small fraction (0.2 per cent) of oxygen atoms. In 1941 four American scientists, Ruben, Randell, Karnen, and Hyde grew the microscopic plant *Chlorella* (*figure 141*) in water which had been treated so as to increase the proportion of $H_2^{18}O$ from 0.2 per cent to 0.85 per cent. Then they examined the oxygen produced by *Chlorella* and found that here, too, the percentage of heavy oxygen averaged 0.85 per cent, rather than 0.2 per cent. They concluded that the *Chlorella* had split up the water molecules and given off the oxygen from them.

If this is so, then we can try to write equations for the processes as if they were simple chemical changes:
1 Water molecules are split into oxygen and hydrogen
   $2H_2O \rightarrow 4H + O_2$
2 Carbon dioxide molecules are reduced by this hydrogen
   $CO_2 + 4H \rightarrow (CH_2O) + H_2O$
   (CH_2O) stands for a simplified carbon compound reduced to the same extent as in glucose. The two equations can be written together

$$2H_2O \rightarrow 4H + O_2 \uparrow$$
$$CO_2 + 4\overset{\downarrow}{H} \longrightarrow (CH_2O) + H_2O$$

11.41 Photosynthesis

The reduction of carbon dioxide to sugars involves the building up of a larger molecule from smaller molecules. Such a chemical change is called a *synthesis* (Greek: putting together). You know from the experiments that sugar formation takes place only in the light. Therefore the term *photosynthesis* is applied to this process. (Greek: *photos* = light).

11.42 Photosynthesis and respiration

Photosynthesis involves the reduction of carbon dioxide to sugars, whereas respiration involves the oxidation of sugars to carbon dioxide (see Chapter 6).

Q1 In how many respects do these two processes seem to be opposite to each other? Draw up a table which will summarize and compare respiration and photosynthesis.

Q2 What is the result if photosynthesis and respiration go on at the same speed in a leaf?

Q3 What must the relationship between the two processes be if a plant is to grow?

In oxidation, energy is released. In the reduction of carbon dioxide to sugars, energy must be taken in. Plants must obtain a source of energy from their surroundings to 'drive' the reduction process.

Q4 From the work you have done what do you suggest as a likely source of energy?

11.5 Chlorophyll and light

Chlorophyll is the substance which makes leaves look green and, like the colouring substance used in paints, it is called a pigment. When you look down at a leaf it appears coloured because light of that colour has been reflected from it towards your eyes. If you hold a leaf up to the light, you will see that it allows light to pass through (colour *plate 6*). The light falling on a leaf from above is white but light emerging beneath is green. To find out why, you can set up an experiment to compare white light with light that has passed through chlorophyll.

1 Prepare a chlorophyll solution by grinding up some green leaves with ethanol or 80 per cent acetone. Filter the mixture to remove leaf remains.

2 Examine white light with a spectroscope. This instrument has a prism or diffraction grating at one end which splits up light into a spectrum of its constituent colours (colour *plate 7*).

3 As soon as you have a clear spectrum put some chlorophyll

Figure 147
Examining a spectrum.

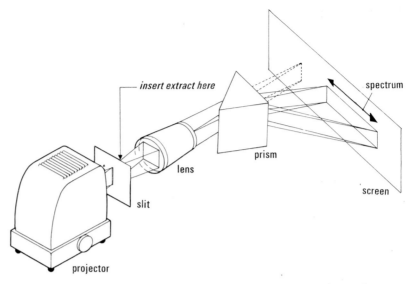

solution in a flat-sided glass or plastic container and lower
it slowly into the path of the light.

Q1 How is the spectrum altered by passing the light through
chlorophyll?

Q2 A spectrum such as the one produced by light which has
passed through chlorophyll solution is called an
absorption spectrum (colour *plate 8*). What has been
absorbed?

Q3 How do you suggest the plant gets the energy needed to
split the water molecule and so reduce carbon dioxide to
sugar?

In section 11.2 it is suggested that variegated leaves should
be tested for the presence of starch. Variegated leaves are
useful experimentally, for they provide the means of
seeing how the non-green plants differ from the green ones.
If a plant used the green chlorophyll to absorb light, can
non-green plants take up carbon dioxide and manufacture
starch when illuminated? You may already know the
answer to part of this from section 11.2.

If you wish to test the carbon dioxide balance of such
leaves you can try putting *a* a normal leaf and
b, a variegated one in corked test-tubes with a little
indicator at the bottom and arrange them side by side in
bright light (see section 10.11).

Then consider the results from the two related
experiments, this one and the results of testing a
variegated leaf for starch (section 11.2).

Q4 Do they support the hypothesis that the chlorophyll in leaves absorbs light energy which is used in photosynthesis?

Figure 148 summarizes the events of photosynthesis.

Figure 148
An outline of the changes taking place in photosynthesis.

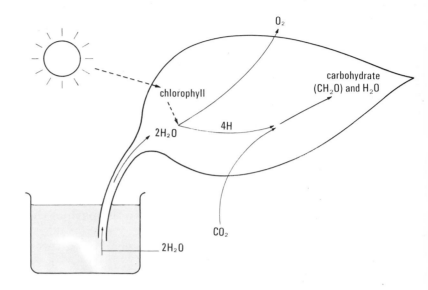

11.6 The structure of leaves

Leaves are used in many of the investigations in this chapter and in Chapter 10 and this must suggest to you that their structure is important for photosynthesis.

11.61 Designing a leaf

From the knowledge you now have of what is required for the process of photosynthesis to take place it should be possible for you to make suggestions for an 'ideal' leaf.

Q1 Consider what has to be available for the process of photosynthesis to take place. What features do you think the structure of a leaf should have, if it is to achieve its functions? Think of both the surface and the internal structure.

You can examine the detailed structure of a leaf to see how it conforms with your own ideas looking first at the outer layers and then at what lies between them.

11.62 The surface of a leaf

Before you look at the surface of a leaf there is one simple experiment you can do.
1 Boil some water in a beaker. Turn off the Bunsen burner and wait for the bubbles to stop appearing.
2 Take a fresh leaf and, holding it by the stalk, put the leaf blade in the hot water.
3 Examine both the upper and lower surfaces of the leaf while it is in the water for about half a minute.

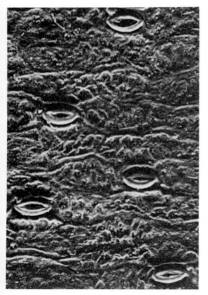

Form a hypothesis to account for your observations. You can examine the surface of a leaf in two ways:
By stripping off the outer layer (*epidermis*).
By making an impression of the surface.

Exercise 1
Take a leaf such as a lettuce or privet leaf and roughly tear it across. It will usually be possible to take hold of the epidermis at the torn edge with a pair of forceps and pull it off the rest of the leaf. A piece of upper and lower epidermis should be taken off, mounted in water on a microscope slide, and examined under a microscope.

Exercise 2
Paint the upper and lower surfaces of a leaf with a thin layer of clear nail varnish or clear adhesive. Allow it to dry thoroughly and then peel it off with a pair of forceps. Put a piece from each surface on a microscope slide in a drop of water and examine under a microscope.

Q2   What differences can you see between the upper and lower surfaces of the leaf you used? Make drawings.

Q3   Do your observations on the epidermis of a leaf confirm or contradict your hypothesis? Give your reasons.

a

wax

cuticular ledge

inner shallow ledge

guard cell

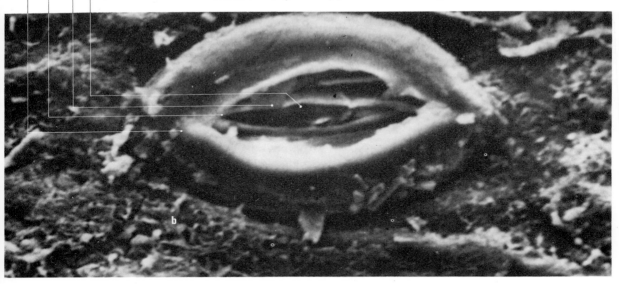

Figure 149
a Stomata on a leaf of *Aloe tenuior var. rubriflora*, × 200.
b A stoma from a leaf of the same plant, × 2750. Note that the pore itself cannot be seen.
Photographs prepared by Dr D. F. Cutler at the Jodrell Laboratory, Royal Botanic Gardens, Kew. Crown copyright.

The pairs of sausage-shaped cells which are in the epidermis of a leaf are called *guard cells*. The guard cells and the pore between them are called a *stoma* (plural *stomata*). You can see several stomata in *figure 149a*. *Figure 149b* is a highly magnified photograph of a single stoma, which may be rather different in appearance from those of the leaves you have examined. *Figure 150* shows what a stoma is like when seen as a solid object.

Figure 150
A stereogram of a single stoma, shown cut in half.

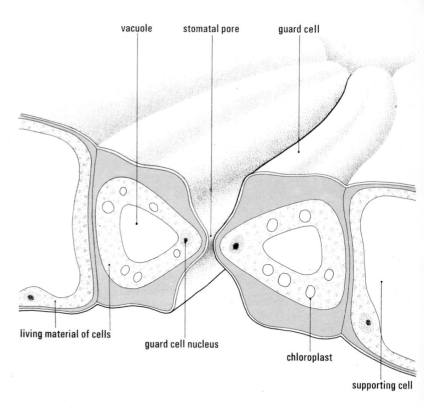

The number of stomata per area of leaf varies. Species of *Tradescantia* have approximately 14 stomata per mm², while Spanish oak has about 1200 per mm². A common number is about 300 stomata per mm² of lower leaf surface.

11.63 The internal structure of a leaf

To find out about the structure of the part of the leaf between the upper and lower epidermis you should look at thinly cut transverse sections of a leaf. These can be either prepared slides or freshly cut sections. In either case examine the section under the low and high power magnification of a microscope.

1   Find the lower and upper epidermis and compare this view with that obtained from a varnish impression or peel of epidermis. Notice what can be seen of the guard cells and stomatal pore (see *figure 150*).

2   If you are using fresh sections, see which parts of the leaf are green and, within any green cells, whether the colour is spread throughout the cell or restricted to particular parts. Compare the sections you look at with the drawings in *figures 151* and *153* and the photograph in *figure 152*, and identify all the parts of a leaf.

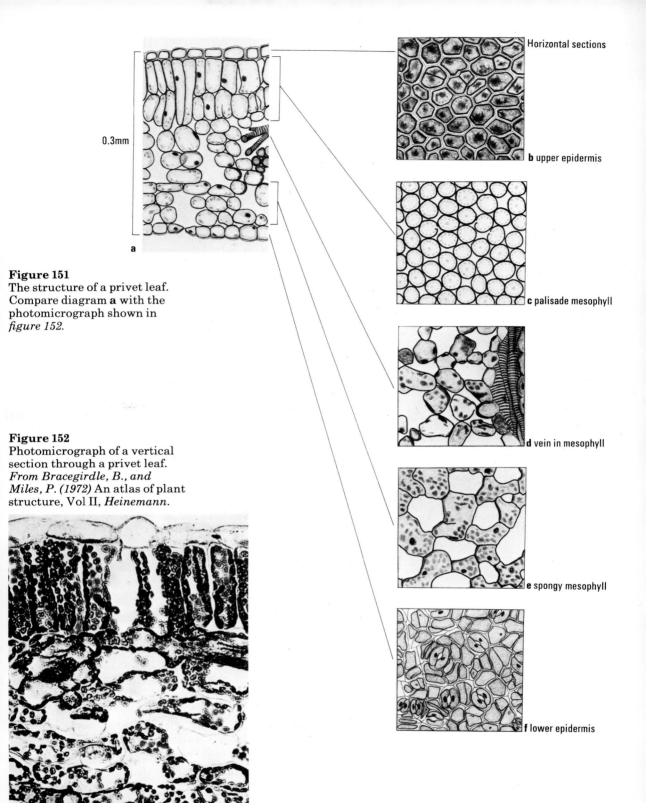

0.3mm

a

Horizontal sections

b upper epidermis

c palisade mesophyll

d vein in mesophyll

e spongy mesophyll

f lower epidermis

Figure 151
The structure of a privet leaf.
Compare diagram **a** with the
photomicrograph shown in
figure 152.

Figure 152
Photomicrograph of a vertical
section through a privet leaf.
*From Bracegirdle, B., and
Miles, P. (1972)* An atlas of plant
structure, Vol II, *Heinemann.*

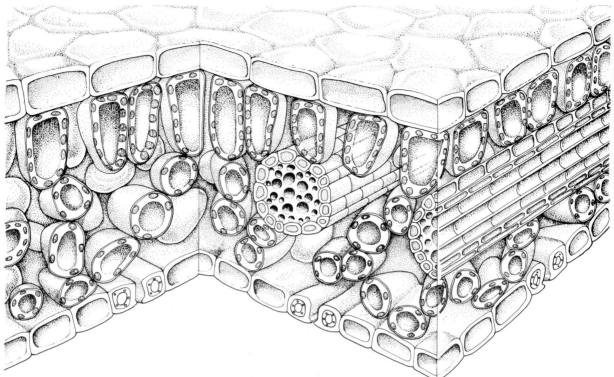

Figure 153
A stereogram of leaf structure.
You can identify the parts that
are shown in *figure 151*, bearing
in mind that here they are
depicted in three dimensions.
*After Gabb, M., and Chinery, M.
(1966)* The world of plants
(Foundations of botany),
Sampson Low, Marston & Co.

Figure 153 shows a stereogram of leaf structure. This
shows a definite upper and lower epidermis together with a
main vein through the centre. Thus, you can see all
features that you examined by looking at surfaces and
between them. This figure and *figure 150* are called
stereograms because they show in three dimensions what
you previously saw in two dimensions only.

11.64 Where in the leaf does photosynthesis take place?

If you examined freshly cut leaf sections you will have seen
small green disc-like bodies in many of the cells – these are
called *chloroplasts* and they contain chlorophyll which
absorbs light (section 11.5). Notice which cells in a leaf
contain them. They are able to change their position
according to the light intensity.

If fresh leaves are ground up in a suitable way the
chloroplasts can be recovered intact and studied apart
from the rest of the cells in which they grow. Scientists
have been able to show that these isolated chloroplasts are
able to carry out the essential process of photosynthesis.
That is, when supplied with carbon dioxide and light

energy, they can manufacture sugars and release gaseous oxygen to their surroundings.

You should now be in a position to consider the close relationship between the structure of a green leaf and the important processes it carries out. Note how the thinness of the leaf brings enormous numbers of chloroplasts within a very short distance of the upper epidermis, which is itself translucent, permitting light to pass through it. The chloroplast will first use the energy from the light to split water molecules and the oxygen gas released will pass out of the cell through the intercellular spaces in the mesophyll layer and thence through the stomata to the atmosphere. Notice the short distance from the centre to the outside of a leaf (you can calculate the exact distance from *figure 151*). Oxygen molecules inside the leaf move outwards by the process of *diffusion*. This is because the chloroplast is constantly releasing oxygen and its concentration inside the intercellular spaces becomes greater than that outside, giving a gradient of oxygen concentration.

The hydrogen split from the water molecules is now used in the chloroplasts to reduce carbon dioxide to sugars. This means that the concentration of carbon dioxide next to the mesophyll cells will be below the normal value in air of 0.03 per cent, and, with the higher concentration in the atmosphere, another gradient is produced, this time causing carbon dioxide to diffuse into the leaf. The sugars made in the cells have to be carried away from them and you would probably suggest that the veins could do this. Notice where the veins are in relation to mesophyll cells, in *figure 151*.

Thus you can see that there appears to be a close relationship between the structure of a green leaf and the important functions it carries out.

11.7 Food webs and energy

This chapter began by looking at the position of green plants in food webs; you can now re-examine this in the light of your knowledge of photosynthesis.

One square metre of ground in temperate latitudes receives about a million kJ (250000 kcal) of light energy per year.

If all of this were stored by plants in photosynthesis, the food produced would contain about a quarter of a man's annual energy requirements at 11 400 kJ (2740 kcal) per day. Unfortunately, plants are not 100 per cent efficient at absorbing and using light energy, as can be seen from *figure 154*.

Figure 154
The fate of sunlight energy falling on pasture.

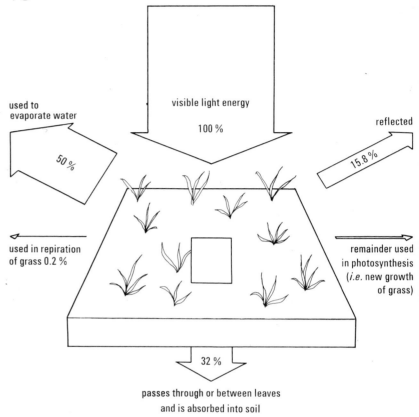

used to evaporate water

visible light energy

100 %

reflected

50 %

15.8 %

used in repiration of grass 0.2 %

remainder used in photosynthesis (*i.e.* new growth of grass)

32 %

passes through or between leaves and is absorbed into soil

Q1   What percentage of the total light energy is incorporated as new plant growth?

Now study *figure 155* which shows how efficiently grass is turned into bullock.

Q2 · What percentage of grass energy becomes beef energy?

The data given in *figure 154* refer to well managed British pasture grassland, but some crop plants are much more efficient than others at trapping and converting energy. A sugar cane plantation may trap 3 per cent during the whole year and a sugar beet crop has been shown to trap 8.9 per cent over short periods. Research into improving crop yields and harvesting the energy over greater areas of the Earth's surface must form an essential part of man's

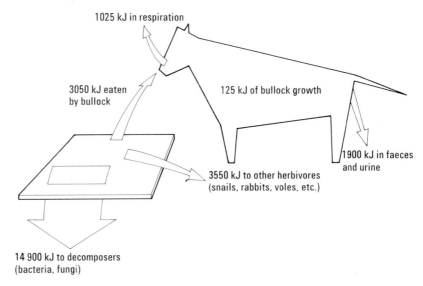

Figure 155
One square metre of British pasture produces 21 500 kJ of new growth a year. What happens to it all?

1025 kJ in respiration

3050 kJ eaten by bullock

125 kJ of bullock growth

1900 kJ in faeces and urine

3550 kJ to other herbivores (snails, rabbits, voles, etc.)

14 900 kJ to decomposers (bacteria, fungi)

activities during the next 100 years if the population explosion is not to have disastrous results. This whole subject is considered in more detail in *Living things and their environment*, Chapter 6.

Background reading

Carbon dioxide for glasshouse crops

Assimilation of carbon dioxide by photosynthesis is the basic process that enables green plants to grow. For some sixty years scientists have been experimenting to see whether they can increase the rate of growth of crop plants by increasing the concentration of carbon dioxide in the air around them. Obviously this is economical in enclosed areas from which the gas cannot escape, and where concentration is easy to control. The addition of extra carbon dioxide to glasshouse crops as a stimulant is now quite widespread.

Normally the respiration of micro-organisms in the kinds of soil used in glasshouses produces carbon dioxide at a rate of from 2 to 3 kg per hectare each hour. If, however, you put manure on a glasshouse bed, it will produce from 20 to 30 kg per hectare each hour – an indication of the rapid rate of respiration of the micro-organisms present in it. When the light in the glasshouse is intense, photosynthesis takes place rapidly and the crop plants take up carbon dioxide rapidly too, keeping pace with its production by the manure and soil. However, in dim light the plants photosynthesize far less quickly, so that the carbon dioxide in the air increases rapidly in concentration and may reach 0.1 per cent – about three times the value in the air outside. Thus the enrichment of the glasshouse air has been carried out unintentionally for many years. How

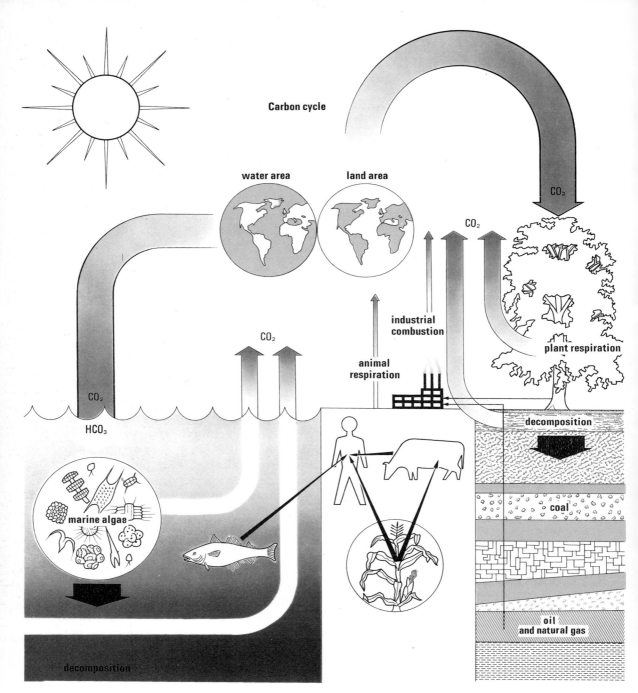

Carbon cycle

water area   land area

CO_2

CO_2

industrial
combustion

animal
respiration

CO_2

CO_2

HCO_3

plant respiration

decomposition

marine algae

coal

oil
and natural gas

decomposition

Figure 156
The carbon cycle. Study this
diagram to get the complete
picture.
*Based on Rabinowitch, E. I.
'Photosynthesis'. Copyright* ©
1948 Scientific American *Inc.;
all rights reserved.*

could the production of carbon dioxide be regulated more
closely? One possible source of the gas was from the
burning of fuels containing carbon – such as the kerosene
which many growers were already using in stoves to heat
the glasshouses. Unfortunately, when many grades of
kerosene are burned, they produce small quantities of the

Figure 157
Two batches of lettuce leaves grown in similar soil for the same time. The batch of larger lettuces, above, had received air enriched with carbon dioxide.
Photographs, Glasshouse Crops Research Institute.

gas, sulphur dioxide, of which even as little as 0.05 per cent has been shown, for instance, to be poisonous to tomatoes. Therefore, care has to be taken to select especially pure kerosene which has very little sulphur in it. Propane, on the other hand, yields carbon dioxide with hardly any sulphur dioxide in it, so many growers are now using this 'bottled' gas as a fuel.

The disadvantage of preparing carbon dioxide in this way is that heat is inevitably produced at the same time. This is useful in cold weather, but in summer the glasshouses get too hot. One solution to the problem is to use solid carbon dioxide as a source of the gas in hot weather; this method produces no poisonous side-effects, but it is more expensive than just burning a fuel. Alternatively, the propane burners can be put outside the glasshouses, and the gases from them can be piped in plastic tubes to the staging on which the plants are grown. How does a grower tell when the concentration of carbon dioxide inside the glasshouse has reached a satisfactory level? The simplest way is to use a modified version of the bicarbonate/indicator solution which you have used to detect changes in the concentration of carbon dioxide in the air. For half an hour 10 cm^3 of the indicator are exposed to the glasshouse air in jars such as honey jars, and then poured into a glass cell with parallel walls so that the colour of the indicator can be compared with a set of standard glass discs. The changes in colour are not very great, and care has to be taken to examine the cell under standard conditions of lighting. Even so, an experienced person can judge the atmospheric concentration of carbon dioxide in a matter of a few seconds, and this means that for purposes of research into plant growth, samples of indicator can be exposed at many points throughout a glasshouse and the variation of the atmospheric carbon dioxide concentration can be followed closely. The next step is to devise an instrument which will control the rate of production of carbon dioxide in response to changes in its concentration – just as a thermostat on a hot water tank switches the heat on and off as the water temperature varies. A simple instrument has now been made which does this cheaply: as more carbon dioxide appears in the air, more will dissolve in some deionized water exposed to it. This causes a change in the electrical resistance of the water – the more carbon dioxide dissolving, the lower the electrical resistance – so that a simple electrical device can be built to decrease the flow of propane to the burner when the atmospheric concentration of carbon dioxide reaches the required level.

What is the 'best' level at which to aim? This is a difficult question to answer simply. Research on the photosynthetic rates of single leaves of crop plants suggests that a concentration of about 0.1 per cent of carbon dioxide would give a considerable increase in the rate of growth; higher concentrations than this would probably be wasteful, as the temperature might be too low and there would be insufficient light energy to enable the plants to use the extra carbon dioxide. But so far there is little comparable information about the requirements of whole plants, and further research is needed. We are still only on the threshold of greater efficiency in food production.

Water and life

12.1 Organisms contain water   The water content of some organisms is shown in *figure 158*.

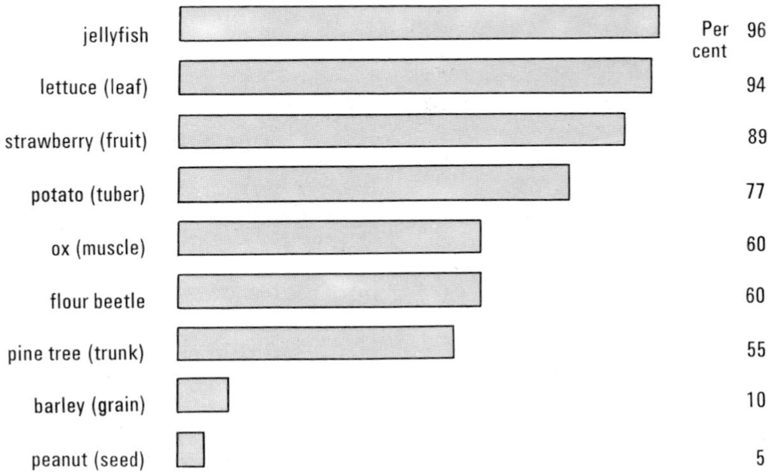

	Per cent
jellyfish	96
lettuce (leaf)	94
strawberry (fruit)	89
potato (tuber)	77
ox (muscle)	60
flour beetle	60
pine tree (trunk)	55
barley (grain)	10
peanut (seed)	5

Figure 158
The percentage of water in
various organisms.

Human beings, like most animals and plants, contain a
high percentage of water. It is rather humiliating to be
largely comprised of so simple a compound – the famous
nineteenth-century biologist Thomas Henry Huxley once
said 'Even the Archbishop of Canterbury is 60 per cent
water!' But perhaps we ought to look into what special
part it plays in the functioning of living material.

12.2 Uptake and output of water in plants

The apparatus shown in *figure 159*, called a 'weight'
potometer (Latin: *potare* = to drink), will help you to
answer the question 'What does a plant do with water?'
The experiment needs a day or two to work.

It will be interesting to compare the results of different
groups, and see how far they agree.

Q1   Every group may not have exactly the same results. What
are the causes of variation in this experiment?

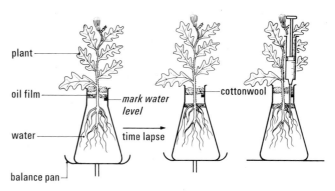

plant

oil film

mark water level

cottonwool

water

time lapse

balance pan

a *Record original mass of plant plus container* b *Note change in mass* c *Add water until water level is restored; note volume to be added*

Figure 159 (*above*)
Using a 'weight' potometer.

Figure 160 (*below*)
A simple bubble potometer.

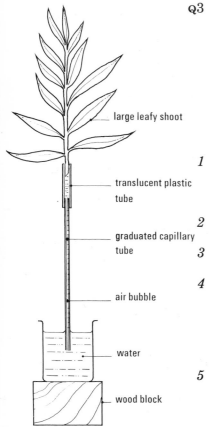

large leafy shoot

translucent plastic tube

graduated capillary tube

air bubble

water

wood block

Q2 What do the results show about what happens to the water taken up by a plant?

Q3 What hypothesis can you suggest to explain the results? How would you test it?

A 'weight' potometer detects the process of water loss which is called *transpiration*. But it must be left for a long time to show any appreciable loss of mass. A bubble potometer can measure how fast a plant absorbs water over a much shorter period of time. A simple bubble potometer is shown in *figure 160* and this is how you use it.

1 Select a suitable plant: this should have a long stem, which is circular in cross-section and has a smooth surface with at least 2 cm of stem between each leaf. Willow, rosebay willowherb, and Michaelmas daisy are examples.

2 Select a piece of plastic tubing about 5 cm long whose internal diameter is slightly less than the stem used.

3 Fix onto the tubing a length of about 30 cm of 0.5 mm bore capillary tubing.

4 Have a bucket of water ready. Choose the stem to be used. With a sharp knife, cut the stem so that the severed portion is several cm longer than required. Plunge it immediately into the bucket and trim off about 5 cm from its base. This avoids introducing air into the stem. Make a slanting cut, to make it easier to fit the stem into the tube.

5 Place the plastic and capillary tubes into a sink or trough of water, ensuring that no air remains in them. The cut ends of the shoots must also be in the same container. Avoid wetting the leaves.

6 Slide the plastic tube gently over the cut end of the stem. It should not tear the 'skin' of the stem. To be quite sure of a firm union with the stem, you can twist a piece of thin copper wire gently round the connection. Leave the union and tube below water for the moment.

7 Set up a stand and clamp, with a block and beaker as in *figure 160*. It is best if the water used has been boiled and allowed to cool. This expels dissolved air, which can cause trouble by appearing as bubbles of gas in the plastic tube and interfering with the flow of water.

8 Put a finger over the open end of the capillary tube and transfer the shoot and tube to the beaker as shown in *figure 160*.

9 Gently blot off any drops of water from the leaves and allow the apparatus to settle for a few minutes. Fasten a scale to the capillary tube or simply mark off two widely separated points (say, 5 cm apart).

10 Lower the beaker until the capillary tube exit is above water level. This lets in an air bubble.

11 Replace the beaker as soon as the bubble is longer than its width.

12 Time the run of the bubble between two marks on the scale.

13 As soon as the bubble has passed the upper mark, pinch the plastic tube gently and expel the bubble.

14 Repeat the run to see how steady the uptake is. Consistent timing over three (or more) runs is a fair indication of this.

Design and carry out an investigation, using a bubble potometer, to find out under which conditions the uptake of water is faster and under which it is slower. Other materials can be used to help carry out the investigation.

Q4 The bubble potometer measures the rate at which a shoot absorbs water. Can you say if this is likely to be the same as the rate at which the shoot loses water (*i.e.* as its rate of transpiration)?

12.21 Transpiration and evaporation

A woman who has a load of washing to hang on the line in cold or overcast weather may complain that it is a 'bad drying day'. You can measure the evaporating power of the air with an atmometer; the one shown in *figure 161* is very simple but works well. Transpiration and evaporation are similarly affected by most environmental conditions, with one exception. Can you think what it is? The leaf's response to this is related to its structure; *figure 162* will help you to decide. When the leaf photosynthesizes, carbon dioxide is used up. This gas passes in through the stomata; meanwhile the oxygen produced passes out. The leaf's

Figure 161
An atmometer. The rubber tubing
has been used to suck up water
into the apparatus.

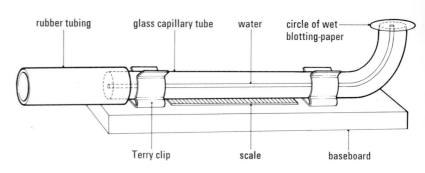

mesophyll tissue has a very large area of cell surface
exposed so that carbon dioxide will be absorbed efficiently.
Such an arrangement must also involve the loss of water
from the cells into the leaf's air spaces. Water vapour from
the humid spaces within the leaf therefore passes out
through the stomata into the open air. Hence,
transpiration is an unavoidable consequence of the leaf's
structure.

The photomicrographs in *figure 162* show how the stomata
of a plant appeared at different stages in a 24-hour cycle.

Figure 162
Stomata on the lower epidermis
of a *Commelina communis* leaf at
different times during a
24-hour period.
*Photomicrographs by courtesy of
Dr T. A. Mansfield, University
of Lancaster.*

Q5 Which of the photographs do you think shows stomata in
the middle of the night?

Q6 Why must a plant have its stomata open most of the time?

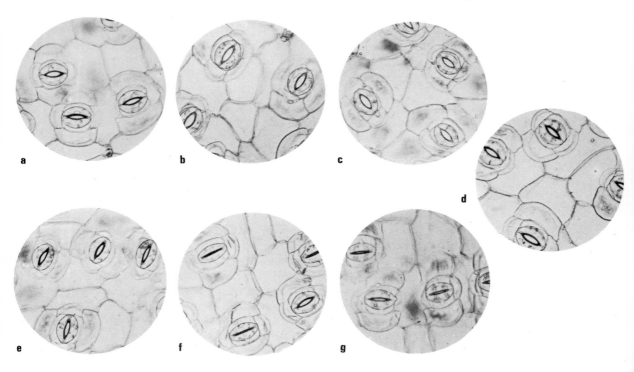

12.3 Movement of water

This experiment needs special cellulose tubing (Visking tubing) of the type used in section 9.1.

1   Take a piece of Visking tubing about the same length as the test-tube or specimen tube you are to use.
2   If the tubing is new, moisten it under the tap and open it out.
3   Firmly tie a knot at one end. Fill the tubing with water and hold it for a minute, checking that no liquid can leak through the knot.
4   If the knot proves to be leaky, discard the tubing or tie a fresh knot above the first and test it again.
5   Empty the water from the Visking tube and, using a bulb pipette or syringe, carefully half-fill it with sugar solution.
6   Push a pin through the top of your Visking tube so that it can be rested across the mouth of the glass tube.
7   Using a pipette or syringe, quickly and carefully add water to the specimen tube until the level is exactly the same as that of the sugar solution inside the Visking tube.
8   You have now set up tube A in *figure 163*.
9   Use a similar procedure to set up tubes B and C. Notice that the water inside the Visking tube in C is a few mm *above* the sugar solution outside it.
10   Leave the tubes for a few minutes and then see if the levels have changed in any of them. Compare your results with those of others.

Figure 163
An experiment to investigate the movement of water. Work quickly in setting up the three tubes, but make sure that the levels start as shown.

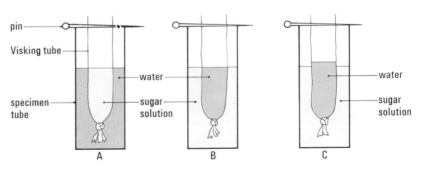

Q1   What has happened to the levels in tube A?

Q2   Tube B should show the opposite result; does it? Can you explain your result?

Q3   Do the results in tube C help you to draw further conclusions?

12.31 Membranes

A membrane is simply a thin layer. There are many such structures in living matter, and they may vary in their ability to let things through. A membrane which freely allows a substance to pass through is said to be *permeable* to it, but if it does not allow that substance to get through it is *impermeable* to it. Some membranes allow many different substances to pass through, but some faster than others; such membranes are said to be *differentially* (or *partially*) *permeable*.

Q4 Looking at your results in section 12.3 what is the simplest hypothesis you can make about the permeability of Visking tubing to water and to sugar?

Q5 How would you test this hypothesis?

Figure 164
An electronmicrograph (× 30 000) of the surface of Visking tubing. *Photograph, R. K. White, Department of Biophysics, University of Leeds.*

Figure 164 shows a piece of Visking tubing very highly magnified. Its surface is evidently not smooth, but one cannot see whether it is porous or not.

12.32 Movement through membranes

When water passes through a differentially permeable membrane from a weak solution to a strong solution we say that *osmosis* has taken place. Osmosis is a very important process in biology, because every cell is surrounded by a membrane that is differentially permeable, all cells contain large amounts of water (see section 12.1), and many cells are surrounded by liquids of various concentrations.

12.4 Osmosis in living cells

Blood is a convenient source of animal cells. The following experiment investigates the effects on red blood cells of three liquids of varying strength.

1 Take three small tubes, each containing about 1 cm^3 of the following fluids:
  a distilled water;
  b 0.85 per cent salt solution;
  c 3.0 per cent salt solution.
2 Transfer a drop of the blood provided to each of the tubes.
3 Shake the tubes gently and leave them for 2 or 3 minutes.
4 Shake the tubes again and, using a bulb pipette, withdraw a drop of liquid from each tube in turn, put it on a slide with a coverslip and examine it under high power. Where blood cells can be seen, make a drawing of a typical cell.
5 Allow the tubes to stand for ten or fifteen minutes.

Figure 165
An experiment to study the effect of liquids of different concentrations on blood cells.

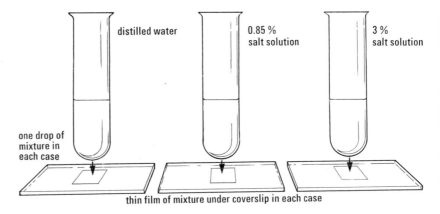

distilled water   0.85 %
salt solution   3 %
salt solution

one drop of mixture in each case

thin film of mixture under coverslip in each case

Q1 What is the appearance of the liquid in the three tubes?

Q2 What do the red blood cells look like under the microscope?

Q3 Explain how the appearance of the liquid in each tube relates to the appearance of the red blood cells as seen under the microscope.

Plant cells in general differ from animal cells in general. *Figure 166* summarizes the main differences.

You can also investigate the effect liquids of different concentrations have on plant cells. One method is shown in *figure 167*. The onion bulb is a good source of living cells, since sheets one cell thick can be peeled from the inside of the bulb scales. Use a red-skinned onion if possible, as the contents of the cell are coloured and thus easier to observe. But an ordinary white onion will do well enough. Note the following precautions:

Figure 166
A typical plant cell and a typical
animal cell.

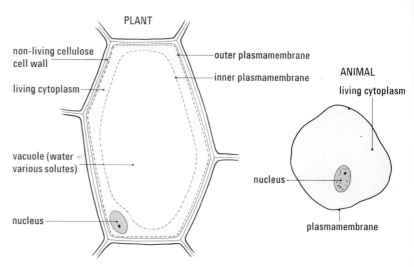

1   Do not let the sheet of cells dry out, or they will die.
    Therefore transfer them to the solution quickly.
2   In using forceps to peel off the cells, remember that any
    cells gripped in the blades will be damaged. Therefore

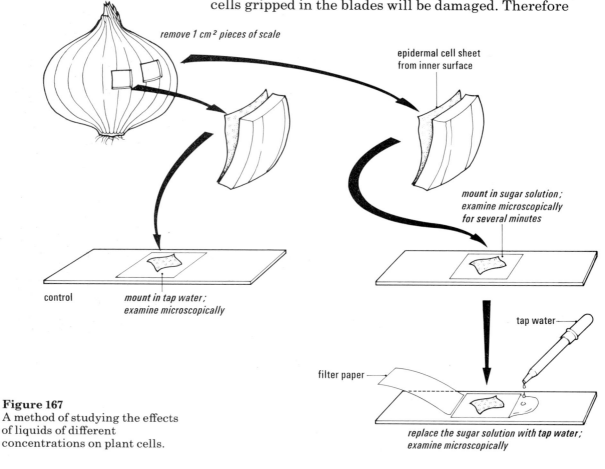

Figure 167
A method of studying the effects
of liquids of different
concentrations on plant cells.

manipulate the cells gently once they are removed (*e.g.* with a brush).

Although onion cells are bigger than blood cells, you will not be able to see all the structures within them. The plasmamembranes are invisible, and in many parts of the cell the layer of cytoplasm is quite thin. Aim to make drawings, of one typical cell of epidermis in each case, mounted in water, mounted in sugar solution, and mounted in water after being in sugar solution.

Q4 In what ways is the action of an onion cell similar to that of a blood cell?

Q5 In what ways do they differ?

Q6 How does the electronmicrograph of a plant cell wall (*figure 168*) help you to understand the reasons for the plant cell's reaction to sugar solution?

Figure 168
An electronmicrograph, showing the surface of a plant cell wall. This is *Chaetomorpha*, an alga; the pattern of its cellulose fibres is similar to that in other plants. Magnification is × 30 000, the same as in *figure 164*.
Photograph, Professor R. D. Preston, Physics Department, University of Leeds.

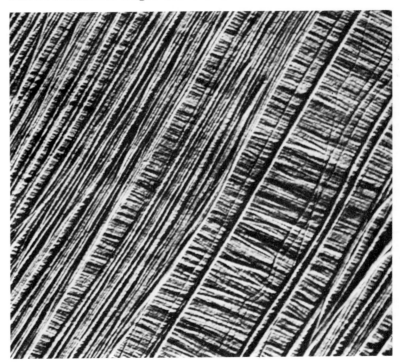

When considering the osmotic behaviour of plant cells, certain specialized terms are useful:
Turgid – the adjective applied to a cell which is fully 'inflated' with water. In such a cell the cytoplasm is being pressed outwards against the cell wall.
Flaccid ('flaksid') – this refers to a cell that is not turgid. In a flaccid cell, the cell wall is not fully stretched and the whole cell will tend to be soft and flabby.

Plasmolysis – the process observed in some of the onion cells, where the cytoplasm shrinks away from the cell wall in places. A plasmolysed cell is a very flaccid one.

Q7 Why does a herbaceous plant usually wilt when its water supply is cut off? Try to use the words turgid, flaccid, and plasmolysis in a single sentence to answer this.

12.5 Osmosis and organisms

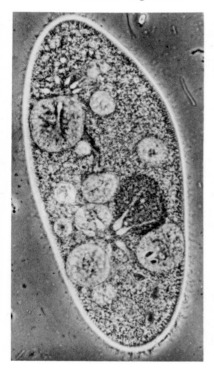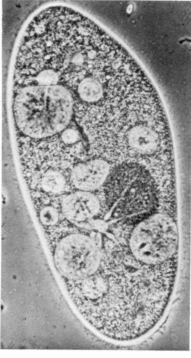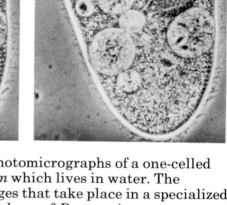

Figure 169
The one-celled pond animal, *Paramecium* (× 600). These photomicrographs were taken at intervals of three seconds. *Photographs, M. I. Walker, by courtesy of Philip Harris Biological Ltd.*

Figure 169 is a series of photomicrographs of a one-celled animal called *Paramecium* which lives in water. The sequence shows the changes that take place in a specialized structure within the cytoplasm of *Paramecium*.

Q1 What does this structure appear to be doing?

Q2 How would its function be related to the animal's habitat?

Q3 Design an experiment which might test your hypothesis.

12.51 Freshwater fish and marine fish

Your ideas in the previous section were probably based on osmosis – the micro-organism tends to absorb water from the pond, stream, or puddle it inhabits. Larger animals, too, have osmotic problems. For a fish, for instance, such problems are very different according to whether it lives in the sea (as the cod does) or in fresh water (as the carp does). You can see in *figure 170* how the cod and carp are adapted to their different environments.

Figure 170
Osmotic problems; a comparison between the cod and carp.

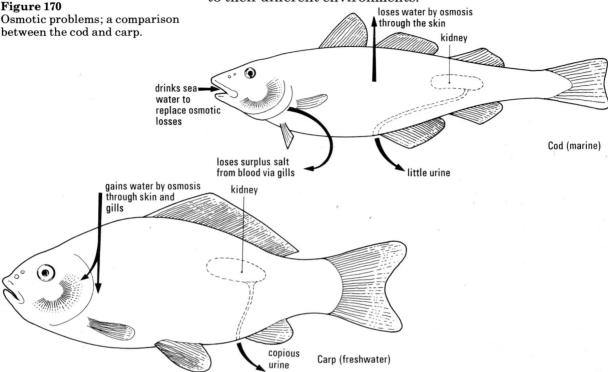

loses water by osmosis through the skin

kidney

drinks sea water to replace osmotic losses

loses surplus salt from blood via gills

little urine

Cod (marine)

gains water by osmosis through skin and gills

kidney

copious urine

Carp (freshwater)

A freshwater fish like a carp produces a large volume of urine; in a day it expels the equivalent of 30 per cent of its body mass.

Q4 How does this compare with human urine output – about 1.5 dm^3 per day?

Q5 Freshwater fish like the carp drink hardly any of the water that they swim about in. Why not?

Q6 What would happen to a cod transferred from the sea to fresh water?

Q7 What would happen to a carp transferred from fresh water to the sea?

In a freshwater fish called the tench, the skin produces a thick, slimy secretion of mucus. If the tench is handled in such a way that the mucous coating is removed, the fish will not survive. On being returned to the water it swells and soon dies.

Q8 What is the function of a tench's mucous coat?

12.52 A rooted plant

The investigations in 12.2 indicated that a plant's roots absorb water. We can now study evidence of how this process works. If some roots are carefully dug up and washed, one can sometimes see *root hairs* that grow out of the region near the tips. These elongated cells grow out into the soil, penetrating the spaces between mineral particles.

Figure 171
Cress roots, showing how the root hairs grow just behind the tip.
Photograph, B. J. F. Haller, Philip Harris Biological Ltd.

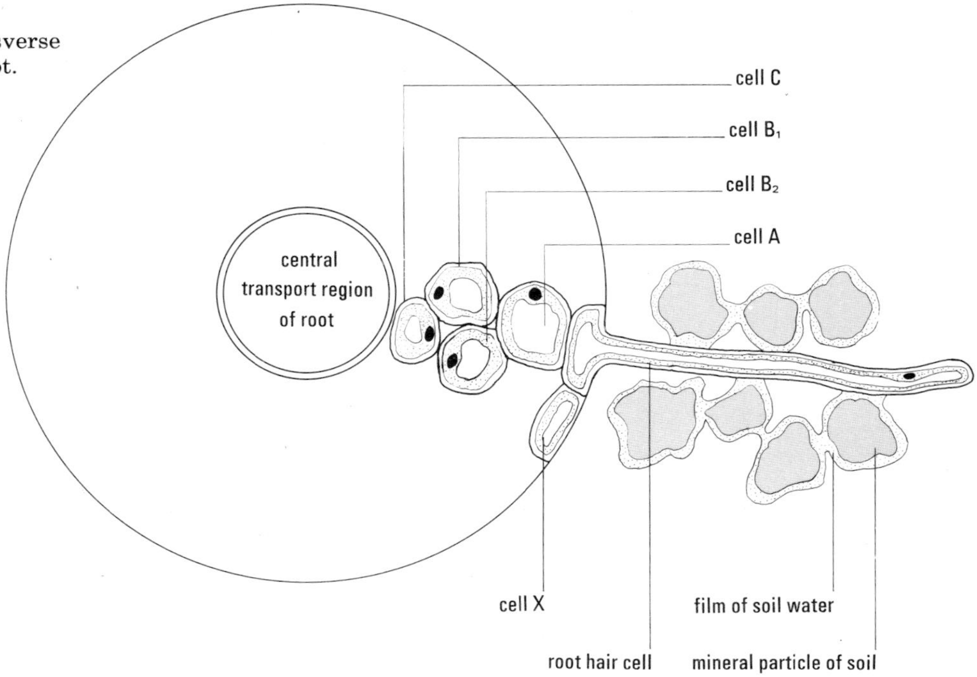

Figure 172
A diagrammatic transverse section of a young root.

cell C

cell B₁

cell B₂

cell A

central transport region of root

cell X

root hair cell

film of soil water

mineral particle of soil

Living things in action

Figure 172 is a very simplified transverse section of a root. You may be able to understand how water passes into the root if you accept the assumption that water is continuously transferred from cell C to the 'transport region' in the middle of the root. No one knows by what mechanism this movement occurs, but all plant physiologists agree that it happens.

Q9   If cell C loses water, what will happen to the cell sap in its vacuole?

Q10   Will this have any effect on cells B_1 and B_2?

Q11   How will any similar changes in cells B_1 and B_2 influence cell A, and the root hair cell?

Q12   Suppose that the cell sap in the vacuole of the root hair cell is a stronger solution than the soil water. What will result from this difference?

When transplanting a small plant it is important to dig it up carefully, leaving the roots still embedded in a ball of soil. If the transplant is carelessly done (for instance, if the plant is pulled up roughly), the plant may wilt in its new place, or even die.

Q13   How may these facts be related to the processes described in this section?

Figure 173
A transverse section through a root's outer layers.
Photomicrograph, W. J. Garnett.

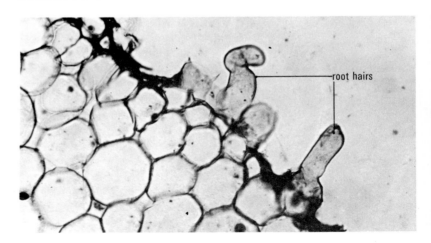

root hairs

Adapt or die

We have seen in this chapter that organisms can face two dangers of water balance; they may absorb too much water or they may lose too much water. We have already seen how there can be barriers to water loss which help to fend off desiccation. These often show beautiful adaptability, as a biologist saw in a vivid way when he spent several months in the desert regions of Mediterranean Africa.

It was summer, and the day temperatures were uncomfortably high. The only sign of vegetation was a sparse, dried-up, thorny scrub, which carried no leaves. At times, when the party halted, the drivers would leave their vehicles and wander across the desert, visiting especially the hollows in the ground, and apparently engaged in the task of picking up small, round, white stones between two and three centimetres in diameter. These they would gather in an old tin can and bring back to their trucks, where they covered them with a pint or two of boiling (and very rusty) water from the radiators. After standing in the water for ten minutes or so, the 'stones' were removed, the water returned to the radiator, and a strange feast began. Each 'stone' was cracked smartly on a rock, when it broke open to reveal a greyish pulpy mass of soft flesh which was eaten with great gusto. The 'stones' were a species of desert snail. It seemed quite incredible that so succulent a creature could tolerate the baking heat of that brutally inhospitable place. But close examination showed that the snails managed this by developing a barrier against their arid environment.

The mouth of each shell was sealed by a thick plate of white, chalky material, presumably containing calcium carbonate like the rest of the shell. Now it would be wrong to say that the animal was hermetically sealed in its shell, for it was alive and presumably respiring, but the plate clearly formed a most effective check to evaporation. In these regions of Africa it usually rains a little in January or February, and during the few weeks of moisture the myriad dormant seeds of many types of plant germinate and carpet the stony ground with their vegetation. The thorny scrub sprouts and the snails erode their sealing plates and go to work feeding and breeding until the supplies of water, and therefore of plant food, give out. The plants set seed and wither. The snails put up their water barriers and enter the long period of summer and autumn inactivity. This is a period of aestivation (Latin: *aestivare* = to spend the summer), the opposite of hibernation (Latin: *hibernare* = to winter). The snail shows adaptability by reacting to a changing environment in such a way as to ensure its survival.

Every animal that lives in very hot conditions shows adaptations which minimize water loss. The camel has a thick skin which is a better water insulator than our own skins. The gerboa, a small desert mammal, moves about mainly at night and spends the day in a burrow – a sort of daily aestivation. Plants that live in deserts are adapted, too; *figure 174* compares the amounts of shoot and root in plants from two very different places. Closer examination of a cactus shows further adaptations which help to conserve water. Perhaps you can think of some of them. Some land organisms have ineffective water barriers and must always live in conditions where evaporation is low. A frog is such an animal. Once it moves out of the damp air associated with grass or a ditch it rapidly runs into trouble; its body fluids diminish in volume, and it dies. The frog is tied, by the ineffectiveness of its skin as a water barrier, to a very specialized and restricted habitat.

Figure 174
Plants from two very different areas. Try to relate each plant's structure to the climate in its area.

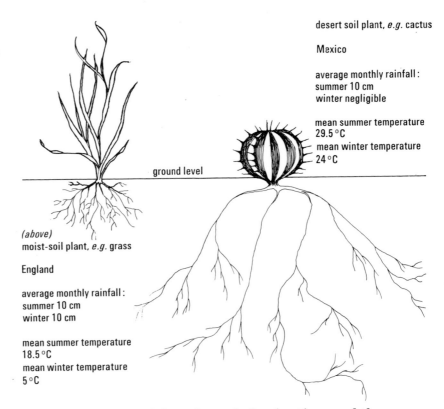

desert soil plant, *e.g.* cactus

Mexico

average monthly rainfall:
summer 10 cm
winter negligible

mean summer temperature
29.5 °C
mean winter temperature
24 °C

ground level

(above)
moist-soil plant, *e.g.* grass

England

average monthly rainfall:
summer 10 cm
winter 10 cm

mean summer temperature
18.5 °C
mean winter temperature
5 °C

The life history of the salmon is fascinating, and there are still many unanswered questions. During the spring and summer months mature salmon enter rivers from the sea. They cease to feed, and make their way higher up the river to the shallow gravelly pools where spawning occurs. Exhausted and emaciated, they drift back seawards. Only a small proportion probably survive to return the following year. Young salmon, known as parr, hatched in the upper reaches of the river, migrate, when a year or more old, towards the sea; then they become known as smolts. For both incoming salmon and outgoing smolt the osmotic changes at the river mouth are severe. In the sea, water tends to be removed from a fish; freshwater surroundings tend to add water to it. The key organs are the gills and kidneys.

For both mature salmon and immature smolt, recent work has shown that structural changes take place in the kidneys during the transition from fresh to salt water and vice versa. The crossing of an estuary is often quite slow, taking several weeks while the fish become acclimatized. But this presents another biological puzzle. Are the kidney changes caused by the migration, are they the cause of migration, or are both migration and kidney change the result of an unknown factor or factors? Modern researchers favour the second view – that a smolt grows a 'sea-going' kidney and so (to quote one expert source) goes to sea 'where it will feel more comfortable'.

You should realize that every animal and every plant faces other hazards than those of water supply – all organisms compete with one another for light, warmth, space, and substances from the environment. In this battle, some species have become specialized for life within a narrow range of environmental conditions (for instance, the dog flea). Other species can cope with wide variability in their surroundings; perhaps man is the best example here, though he often copes with his environment by altering it – air conditioning in the tropics, central heating in cooler zones. (See also colour *plate 2*.) But whether an organism is *adapted* or *adaptable* we can certainly say 'adapt or die'.

Substances in solution

13.1 Solutes and plant growth

Figure 175
Sand dunes at Braunton
Burrows, near Barnstaple, North
Devon. The turfs used in the
experiment were taken from a low
part of the dune system. Its
vegetation was similar to that of
the flat area in the middle distance.
*Photograph by courtesy of
Professor A. J. Willis and the
British Ecological Society; from
J. ecol. (1963) 51.*

Figure 176
The lefthand pot holds a turf
treated for 31 weeks with the
solution. The righthand pot holds
a control turf. Each plot is 28 cm
across.
*Photograph, Professor A. J.
Willis, Department of Botany,
University of Sheffield.*

In a recent investigation, botanists studied the growth of
plants occurring on sand dune soils. They selected an area
on the North Devon coast.

Turfs were carefully removed from the dune system in *figure 175* and taken to the laboratory. One group of turfs was 'watered' with a solution containing these substances:

potassium nitrate (0.202 g) ⎱ per
calcium nitrate (0.656 g) ⎰ dm³
sodium phosphate (0.208 g) ⎱ of
magnesium sulphate (0.369 g) ⎰ solution

Another group of turfs was kept equally moist but not provided with the substances listed above. *Figure 176* shows the results of these two treatments.

Q1   In what ways do the plants in the two pots differ?

Q2   This experiment gives especially clear results because it uses plants from *sand dunes* – why do you think this is?

13.11 Mineral salts in soil and in plants

Table 20 shows the results of analysing the soil in a Derbyshire dale and six plants which grow there.

Plant contents	N	P	K	Ca	Mg	Mn	Zn	Pb	Al
	(mg gram⁻¹ dry mass)					(µg gram⁻¹ dry mass)			
Centaurea nigra (knapweed)	19.7	1.07	17.3	24.0	2.60	100	33	25	199
Lotus corniculatus (birdsfoot trefoil)	26.5	0.89	12.0	35.3	2.30	120	40	13	59
Poterium sanguisorba (salad burnet)	19.00	1.10	10.3	28.7	5.20	120	20	12	119
Festuca ovina (sheep's fescue; a grass)	12.3	0.76	12.7	4.1	0.75	50	25	1	37
Helictotrichon pratense (meadow oat grass)	16.3	0.98	19.3	5.1	1.30	140	9	3	34
Carex flacca (a sedge)	15.7	1.10	14.7	7.5	1.50	150	23	5	66
Soil contents	N	P	K	Ca	Mg	Mn	Zn	Pb	Al
	(mg 100 grams⁻¹ dry mass)					(µg 100 grams⁻¹ dry mass)			
	0.95	0.16	10.8	928.3	18.0	1.6	0.4	0.6	less than 0.1

Table 20
Data provided by Dr. I. H. Rorison.

The bottom line in the table shows how much of each element the soil contains.

1   Arrange these elements in order with the most abundant first and the least abundant last (1 mg = 1000 µg).

2   Make a similar arrangement for one of the plant species; if each person chooses a different species you can pool your results.

Q3   What sort of general pattern emerges from your comparison of plant contents and soil contents?

13.2 Mineral nutrition

Figure 177
Julius von Sachs (1830–97), Professor of Botany at the University of Würzburg; a pioneer in plant nutrition. *By courtesy of The Wellcome Trustees.*

Chapter 11 deals with the chemical reactions by which green plants synthesize sugars and starch. But plants can also manufacture proteins.

Q1   What additional element will be required and from what source might the plant obtain it?

In the early nineteenth century the advance in chemical techniques, especially in Germany, led to increasing knowledge of plant nutrition. The great German botanist Sachs (see *figure 177*) showed that a plant would grow well in the following mixture:

Sachs's culture solution (per 1 dm³)
potassium nitrate (1.0 g)	magnesium sulphate (0.5 g)
sodium chloride (0.5 g)	calcium phosphate (0.5 g)
calcium sulphate (0.5 g)	iron(II) chloride (trace)

A contemporary of Sachs, Wilhelm Knop, discovered that the sodium chloride could be omitted without affecting the plant's growth. He also altered the solution's components, supplying the same elements as different compounds.

Knop's culture solution (per 1 dm³)
calcium nitrate (0.8 g)	magnesium sulphate (0.2 g)
potassium nitrate (0.2 g)	iron(III) phosphate (trace)
potassium dihydrogen phosphate (0.2 g)	

13.21 The plant's mineral requirements

Knop, Sachs, and other investigators of their day devised many different 'recipes' for culture solutions, all of them based on trial and error (empirical methods). You can repeat some of Knop's classic work. You can discover whether a plant needs a particular substance by omitting it from the mixture.

Q2   Suppose you made up some Knop's culture solution but left out the iron(III) phosphate. Which ion or ions would be completely deficient in such a solution?

Q3   Suppose you made up some Knop's culture solution but left out the magnesium sulphate. Would it be correct to call this a magnesium-deficient solution?

Q4   How would you suggest making up some Knop's culture solution which would be deficient in nitrogen?

There are many ways of growing plants in mineral solutions (see *figure 178*). See how well your plants grow with distilled water, complete culture, nitrogen-deficient, magnesium-deficient, and iron-deficient culture solutions.

Figure 178
Different methods of growing plants in mineral culture solutions.

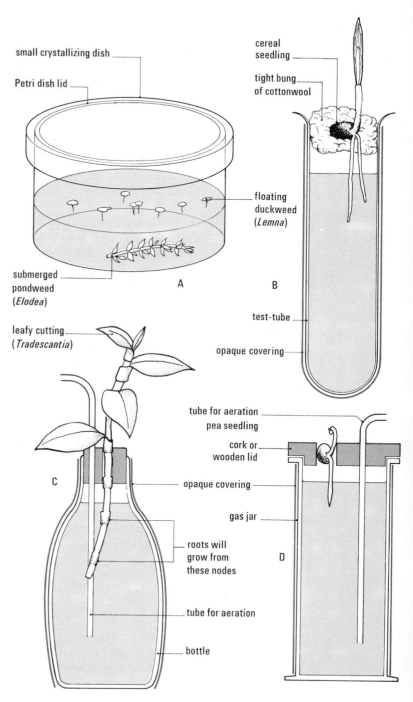

small crystallizing dish

Petri dish lid

floating duckweed (*Lemna*)

submerged pondweed (*Elodea*)

A

cereal seedling

tight bung of cottonwool

test-tube

opaque covering

B

leafy cutting (*Tradescantia*)

C

roots will grow from these nodes

tube for aeration

bottle

tube for aeration
pea seedling

cork or wooden lid

opaque covering

gas jar

D

Whichever method you use, there are some basic points of procedure you should observe:

1 The experiment will show you how mineral supply affects the growth of a plant, so you must decide how to measure this growth. Different methods suit different plants.

Q5 What methods can you suggest which would be a measure of plant growth for your experiment?

2 If you are using gas jars or milk bottles, you should arrange for the culture solution to be aerated for ten minutes every day or so. The air should not be bubbled through too fast, because wetting the bung or lid may make it go mouldy.

3 Every week, top up the culture solution with distilled water to maintain the level.

4 Notice that three of the containers have an opaque cover (black paper, black polythene, or aluminium foil). You can remove this to examine the root growth, but put it back.

Q6 Why is this cover important?

13.3 How are minerals absorbed?

Mineral culture experiments indicate what use a plant makes of particular mineral ions, but shed no light on *how* ions are taken into the plant. You can find out more about the mechanism of absorption from the data in *figure 179*. An American scientist called Hoagland determined the concentration of certain ions in the cell sap of a freshwater alga, also measuring the concentration of those ions in the pond water.

Figure 179
Hoagland's experiment with the alga, *Nitella clavata*.

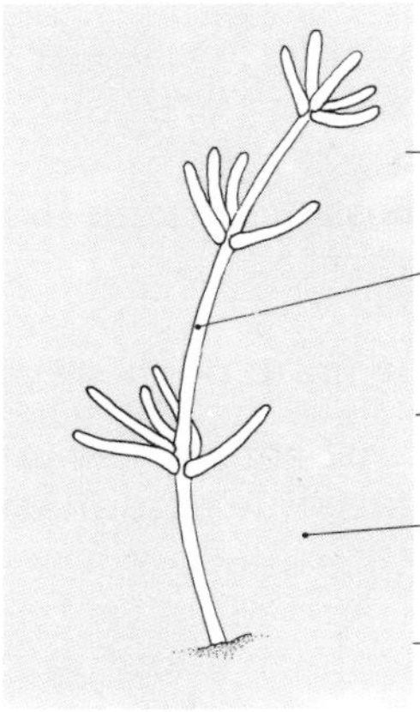

	Na^+	K^+	Mg^{2+}	Ca^{2+}	Cl^-
	(ion concentration in mg/dm³)				
cell sap removed from these cells and analysed	1980	2400	260	380	3750
pond water also analysed	28	2	36	26	35

Q1 How does the concentration of ions in the cell sap compare with their concentration in the pond water?

Q2 Could any of the ions have passed into the plant's cells by diffusion? (See Chapter 14, page 210.)

Many scientists have studied the ion uptake of plants with interesting results. In one experiment barley seedlings were grown in a mineral culture, in which the sulphate ions all contained the radioactive isotope ^{35}S (see Chapter 10, page 153). The 'normal' isotope of sulphur is ^{32}S. Two similar groups of seedlings were used but oxygen was bubbled through the culture solution of one group and nitrogen through the other (see *table 21*).

Table 21
Relative rates of uptake of sulphate ions in aerobic and anaerobic conditions, by barley seedlings.

Time from beginning of experiment (minutes)	Amount of sulphate ions taken up:	
	With oxygen (aerobic)	With nitrogen (anaerobic)
0	0	0
30	220	140
60	290	190
90	350	210
120	390	225
150	430	238
180	470	250
210	500	260
240	530	270

The results were obtained by seeing how much ^{35}S was left in the culture solution at given intervals of time and subtracting it from the known original content of ^{35}S. The difference represents the amount of sulphur as sulphate ions which have gone into the plant.

Plot graphs of the results.

Q3   By what means would the ^{35}S have been detected?

Q4   What conclusions can you draw about the uptake of sulphate ions under aerobic and anaerobic conditions?

Q5   What substance is present in aerobic but not anaerobic conditions?

Q6   What process occurring in most living organisms uses this substance?

Q7   What is the main purpose of this process?

Q8   Using the answers to the previous questions can you suggest a possible explanation for the results given in *table 21*?

Further evidence about ion uptake comes from an experiment in which thin discs of living potato tuber were floated in a solution of bromide. After some time the bromide absorption was measured, by determining the amount left in the solution. Several sets of discs were used, each receiving a different percentage of oxygen in the air bubbled through the solution. *Figure 180* shows the results.

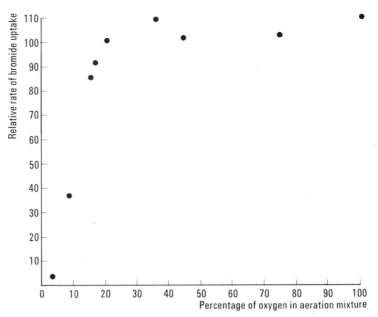

Figure 180
Bromide absorption by living potato tissue.
Data from Steward, F. C. (1959)
Plant physiology: a treatise,
Volume II, Academic Press.

Q9 How does oxygen supply affect bromide uptake?

Q10 In the same experiment it was found that those cells which had absorbed bromide most strongly now contained less starch than other cells. How might this be relevant?

Q11 Potatoes do not usually contain bromide, and will grow quite normally if not supplied with it. Why then was this ion used in the experiment?

Q12 The crop yield of plants growing in waterlogged soils is greatly increased by draining the soil. What do you suggest is one reason for this improvement?

13.4 Transport from the soil

When conditions are favourable for transpiration (see section 12.21), a plant may lose 2.0 grams of water every hour through every 100 cm^2 of leaf surface. In order to make good this loss and to provide water for other purposes, water must be taken from the soil. This water will also contain mineral ions. It follows that the plant must have some means of transporting these materials to the parts of the plant where they are needed. You can trace the pathway of liquid in a plant, using a solution of a harmless dye.

1 You are provided with cut leafy shoots which were placed in a weak dye solution some time before the lesson.
2 Examine the stem, leaf stalks, and undersides of the leaves. Use a hand lens and work in good light from a window or bench lamp.

3   To examine the inner parts of the stem, cut off the part that was in the dye solution with a stiff-backed razor blade.

4   Now cut across the stem at about 2 cm intervals, working downwards (see *figure 181*). Use a lens to examine the cut surface each time, looking out for signs of dye.

Figure 181
Tracing the upward flow of a dye solution.

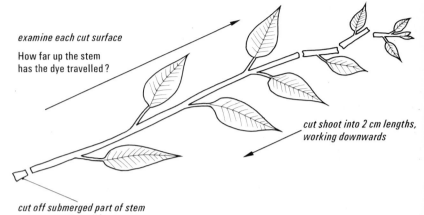

examine each cut surface

How far up the stem has the dye travelled?

cut shoot into 2 cm lengths, working downwards

cut off submerged part of stem

Q1   If you know how long the shoot has been in the dye solution, calculate the rate of upward movement in cm per hour.

5   Use the razor blade to cut very thin slices of stem – thin enough to be transparent – and mount these in water on a slide. Examine with a hand lens or under a microscope.

Figure 182
Photomicrographs of sections of the stem of a broad bean (× 45).
Photographs, W. J. Garnett.

6   The sections should resemble *figure 182* or *183*. Decide which part of the stem has conducted the dye solution.

phloem tubes

xylem tubes

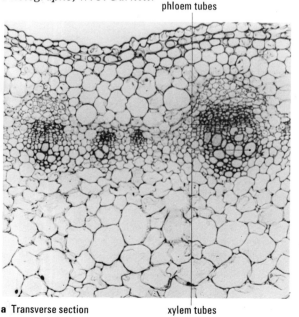

a Transverse section

xylem tubes

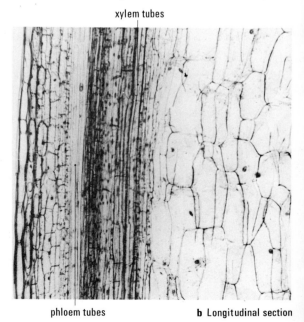

phloem tubes

b Longitudinal section

Living things in action

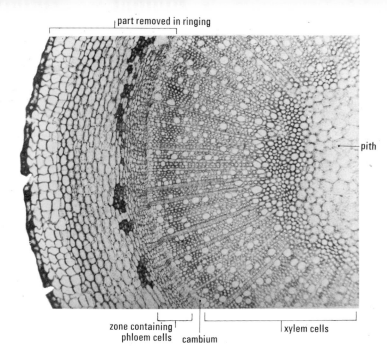

part removed in ringing

pith

zone containing
phloem cells   cambium   xylem cells

Figure 183
Photomicrograph of a transverse
section of privet stem (× 100).
Photograph, W. J. Garnett.

13.5 Transport from the leaves

Experiments with dyed water show how water is delivered
to the leaf. Sugars, and compounds derived from them, are
not only found in the leaf where they are made, so we
should look for a system whereby these substances could
be transported about the plant.

As long ago as the eighteenth century, Stephen Hales
performed some experiments which throw some light on
the problem. He removed a ring of bark and phloem from
the branch of a tree, with the result shown in *figure 184.*

Figure 184
The effect of bark ringing.
After Hales, S. (1727) Vegetable
staticks.

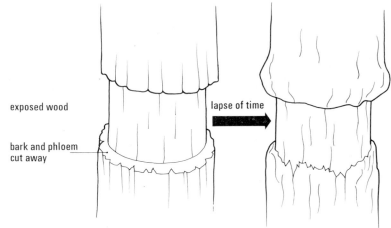

exposed wood

lapse of time

bark and phloem
cut away

Q1   What can you conclude from the change just below the
ring?

Q2   What can you conclude from the change just above the
ring?

Another worker, Malpighi, noted that these changes did not take place if the experiment was performed in winter, when the trees lacked leaves.

Q3 In what way is this observation relevant?

Figure 183 shows the appearance of a woody stem in transverse section, and shows which parts are normally removed by ringing. Any of them might be the site of downward transport.

Q4 From the structure of these parts which would you suggest is most likely to carry out the function of transport?

Nowadays we can employ much more refined techniques, using compounds of ^{14}C as tracers. Such compounds produce an effect on photographic film.
The photographs in *figures 185* and *186* illustrate the beginning and end of an experiment of this kind.

Figure 185
Applying radioactive sucrose solution to a ringed privet shoot and its unringed control. The sugar enters the leaf and its path can be traced.
Photograph, Dr J. Hannay, Department of Botany, Imperial College of Science and Technology.

204

Young shoots of privet were cut under water and transferred to test-tubes containing water. In one group the stems were 'ringed' about four leaves from the apex. You can see this clearly in the righthand specimen of *figure 185*. The bark and phloem tissue had been carefully removed until the xylem was exposed. The control group of shoots was not ringed.

Radioactive sucrose, containing ^{14}C, was introduced into the cells of a leaf by first rubbing a very small area of the upper surface with fine emery paper to remove the waxy cuticle. This was done in both ringed and unringed shoots. The treated leaves were then 'fed' with a drop of ^{14}C sucrose solution and allowed to stand in the laboratory for 24 hours. See *figure 185* (plant A).

They were then washed to remove surplus sucrose from the surface, and the shoots were pressed between blotting-paper and dried by gentle warmth. Each shoot was now placed in close contact with X-ray film in a light-proof folder for five days, at the end of which time the film had become affected in the manner shown in *figure 186*.

Look again at the transverse section of the privet stem as seen microscopically in *figure 183*, so that you can see the location of xylem, phloem, and bark.
Try answering the following questions relating to this experiment:

Figure 186
The shoots have been pressed flat after allowing 24 hours for the radioactive sucrose to disperse. The effect of each leaf on a photographic film is shown in the centre of the picture.
Photograph, Dr J. Hannay, Department of Botany, Imperial College of Science and Technology.

control shoot A   radiographs taken after 24 hours   experimental shoot B

A   B

ringed here

fed leaf

fed leaf

Q5　What evidence is there that radioactive sucrose has been transported away from the leaf into which it was 'fed'?

Q6　What evidence is there that the sucrose can be transported both up and down the stem?

Q7　Is there any evidence that the xylem is responsible for transporting the sucrose?

Q8　How far do the results support the hypothesis that phloem is responsible for transporting the sucrose?

Q9　What hypothesis can you suggest to account for the appearance of radioactive sucrose in the young leaves but not in older ones?

Q10　What other comments can you make on the distribution of radioactive sucrose in control shoot A?

Q11　Suggest a cause for the movement of sucrose out of a leaf into which it has been introduced.

Q12　Why should ringing a tree trunk kill the tree?

One obvious method of obtaining direct evidence of the function of phloem tubes would be to analyse their contents. It is not at all easy to take samples from such minute tubes, but in recent years a most ingenious method has been devised.

Many insects such as greenfly (aphids) suck the juices from soft stems through their hollow, needle-like probosces. The aphid inserts its proboscis into the phloem, which indicates the presence of a rich food source. The insect is killed and the body carefully cut away, leaving the hollow proboscis sticking from the stem like a minute open hypodermic needle. The contents of the phloem tubes exude from the open end of the proboscis and can then be analysed. They are found to be rich in sugars and amino acids. Here is strong evidence that the phloem tubes provide a pathway for the transport of these substances.

Background reading

Muck and mystery – a historical survey

Long ago man changed from a hunter to a farmer and began to use muck or animal manure as 'food' for his crops. This was undoubtedly an early development in agriculture. Manuring is especially important in areas with high population density, where yields must be correspondingly high. In European and Far Eastern countries this has led

Figure 187
Justus von Liebig (1803–73),
Professor of Chemistry at Munich
University.
*By courtesy of The Wellcome
Trustees.*

SIR J. B. LAWES, AGRICULTURAL CHEMIST.

Figure 188
J. B. Lawes (1814–1900), founder
of Rothamsted Experimental
Station. From the *Illustrated
London News* (1882) *80*.
*By courtesy of The Wellcome
Trustees.*

to the practice of *mixed farming*, in which the manure from
cattle or pigs is used to 'dress' or fertilize the land on
which crops are grown. Back in Elizabethan times the
farmer poet Thomas Tusser described the practice in a
nutshell; calling manure 'compas' (compost) he wrote
'One aker well compast is worth akers three.
At harvest thy barn shall declare it to thee.'

As with all early agricultural methods, manuring was not
worked out from theoretical knowledge of soil nutrients
and plant physiology, but from everyday observation,
trial, and error. Such a rule of thumb approach – called an
empirical approach – has now been replaced by our
increased factual knowledge of how manure works. In less
than two centuries, agricultural science has unravelled a
good deal of the mystery by studying the muck.

The Swiss scientist, de Saussure, laid the foundations of
soil chemistry early in the nineteenth century by showing
that, when a plant is burned, its ash contains other
elements than carbon, hydrogen, and oxygen; he rightly
deduced that these elements must have come from the soil.
But the first great soil chemist was Liebig (see *figure 187*);
he made detailed analysis of plant ash, and recognized that
the soil's fertility depended on its supplying the plant with
essential elements – especially calcium, phosphorus, and
potassium. If any one of these was absent, the soil would be
barren.

Yet Liebig's experiments missed the vital element,
nitrogen. Oxides of nitrogen are all gases, so during
burning all the nitrogen compounds pass into the air. It
was John Bennett Lawes (see *figure 188*) and his colleague
Joseph Henry Gilbert who realized that nitrogen
compounds in the soil are important. Liebig believed on
the other hand that plants could get all the nitrogen they
need from the air, as they do with carbon. Lawes owned
large farm estates at Rothamsted, Hertfordshire, and he
founded there the world's first experimental agricultural
station. The famous Broadbalk field was used to resolve the
controversy about nitrogen supply.

The field was divided into many plots, and every year since
1843 wheat has been grown on each plot. *Figure 190* shows
some of the results from the Broadbalk experiment.

Figure 189
An aerial view of the Broadbalk wheat experiment at Rothamsted Experimental Station, Harpenden, Herts. The field's eighteen plots extend up to the top of the picture, with a wide fallow strip running across all of them. *Photograph, Rothamsted Experimental Station.*

The more nitrate supplied, the bigger the crop. Annual applications of nitrogenous compounds replace what is removed in the garnered crop. By contrast, the soil seems able to supply at least some potassium and phosphate, since satisfactory wheat crops have been grown for more than a century on one plot which has received nothing but nitrate. No one would pretend that dosing the soil with nitrate once a year is good agricultural practice; the soil is more than just a source of ions. But the work of Gilbert and Lawes established that plants need certain ions from the soil, and that nitrate ions or phosphate ions are good value to the plant whether they come from manure or from a chemical factory.

Figure 190
Samples of wheat from four of the Broadbalk plots. Plot 3 is a control, receiving no fertilizer and no organic manure. The other plots received phosphorus as superphosphate, potassium, sodium and magnesium as sulphates plus the stated amounts per acre of ammonium sulphate. (1 acre = 0.4 hectare; 200 lb = about 90 kg.) Note how variation in nitrogen supply affects crop yield. *Photograph, Rothamsted Experimental Station.*

The use of chemical fertilizers has grown rapidly during this century. While manure is good for the soil, in particular in maintaining its physical structure, it contains relatively small quantities of minerals. Chemical fertilizers, being pure substances, provide a richer source of essential ions.

The statistics of fertilizer consumption show how farmers have responded to the availability of factory-produced substances.
World fertilizer consumption
in 1900: about 1 000 000 tonnes
in 1968–9: 60 000 000 tonnes

In Britain during 1969 one large chemical company was producing agricultural ammonium nitrate at the rate of 2600 tonnes – 897 tonnes of nitrogen atoms – per day.

Figure 191
World use of fertilizer, 1968–69.

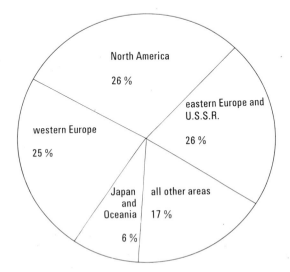

There is some evidence that indiscriminate use of fertilizers may cause long-term harm. Occasionally the minerals may be applied to the soil too liberally and drain away in the rain water; if washed into streams and lakes they may stimulate a sudden growth of algae which in turn upsets the balance of aquatic life. This process, *eutrophication* ('over-feeding'), is a sign that the farmer is wasting his capital outlay by giving his crops more fertilizer than they can use. We may also be neglecting the importance of soil structure. Adding manure or other organic matter to the soil helps it to remain firm but porous, whereas the intensive application of artificial fertilizers can produce a compacted, badly drained soil. This deterioration may be intensified by using heavy machinery to work the soil and by growing the same crop every year.

However, few people doubt that artificial fertilizers will remain an important aid to agriculture. As long as we are fully aware of any dangers through their over-use, these pure and cheaply manufactured substances will go on helping our crops to turn sunlight energy into food substances with maximum efficiency.

Transport in animals

14.1 Movement by diffusion

In order to demonstrate the movement of a substance by *diffusion* you will be provided with a dish in which there is a 2 cm depth of agar jelly.

1 Cut out cubes of jelly in the following sizes:
2.0 cm side 1.0 cm side 0.5 cm side.
One dish will provide enough cubes for three groups.

2 Place the three cubes in a small beaker and cover them with potassium permanganate solution. Leave for five minutes.

3 Pour the permanganate solution off into another beaker.

4 Dry the surface of the cubes thoroughly with blotting-paper.

5 Cut each cube in half and examine the cut surfaces.

Q1 Which of the cubes has been most thoroughly penetrated by the coloured solution and which has been least penetrated by it?

There are other ways of demonstrating the process of diffusion which you may also look at.

Q2 How would you define the process of diffusion?

Q3 How is it similar to or different from the process of osmosis? (See Chapter 12, page 184.)

While no animal is simply a block of jelly, the results of the previous experiment will help you to understand how cells can obtain materials from their surroundings. The same principles apply to the *outward* movement of substances from cell to surroundings.

Q4 Which substances are likely to move into, and out of, cells?

The exchange of materials takes place at the cell's surface, and hence depends on that cell's *surface area*. The cell's needs will be determined by the amount of living material in it – its *volume*. The relationship between surface area and volume is considered in Chapter 3 in connection with heat loss in large and small mammals, where similar principles apply.

14.11 Diffusion and mass flow

Here is a simple experiment which shows the difference between the two ways in which a substance moves.

You will be provided with two glass tubes.

1 Using forceps, place a piece of damp red litmus paper half way along each tube.
2 Place another piece of the damp red litmus paper at one end of each tube.
3 At the opposite end of each tube place a piece of cottonwool which has had a few drops of ammonia solution added. Ammonia has an irritating smell so do not breathe it in directly.
4 Cork one tube at both ends and leave the other open (see *figure 192*).

Figure 192
Investigating diffusion and mass flow.

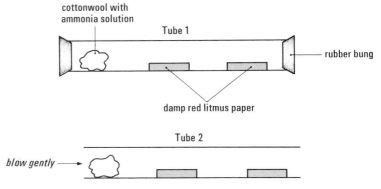

5 Leave tube 1 on the bench, making sure it cannot roll off. Note the time.
6 Blow gently into the end of tube 2 as shown in *figure 192*.

Q5 How is the litmus paper affected in each tube?

Q6 How did the ammonia vapour reach the litmus paper in each tube? One of the methods is diffusion, the other is *mass flow* – which is which?

It would seem from these experiments that distribution of substances in solution by diffusion is only effective for comparatively small organisms. Larger organisms will need two major changes in their structure in order to operate efficiently:

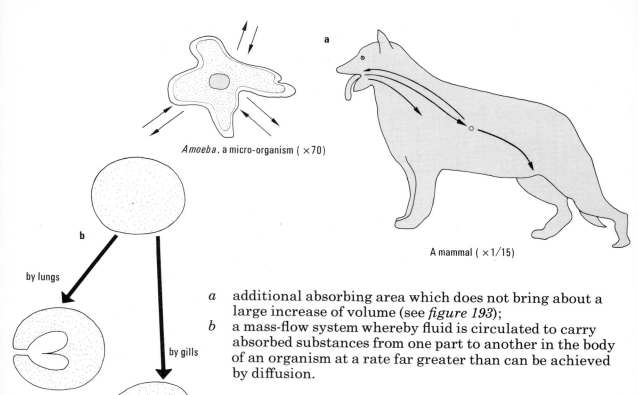

Amoeba, a micro-organism (×70)

A mammal (×1/15)

by lungs

by gills

b

a   additional absorbing area which does not bring about a
large increase of volume (see *figure 193*);
b   a mass-flow system whereby fluid is circulated to carry
absorbed substances from one part to another in the body
of an organism at a rate far greater than can be achieved
by diffusion.

Mass-flow systems increase the efficiency of organisms by
bringing materials rapidly to the right place at the right
time. Our own bodies contain several such systems.

Q7   How many mass-flow transport systems can you identify in
animals or plants?

14.2 Watching blood systems in action

Many small animals without backbones (invertebrates) are
sufficiently translucent for the movement of their blood to
be partly visible. Three of these animals are freshwater
organisms:
the freshwater shrimp *Gammarus* (*figure 194*)
the freshwater louse *Asellus*
the water flea *Daphnia*

Figure 194
A microscope slide and cell with
Gammarus.

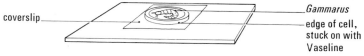

coverslip
Gammarus
edge of cell,
stuck on with
Vaseline

1   Place your specimen in the small cell cut from plastic
tubing and stuck on the slide with Vaseline.
2   Put a drop of water into the cell and place the slide on the
microscope.

Figure 193 *(opposite)*
a Both the animals in this diagram must take in oxygen and food, while getting rid of carbon dioxide and other waste products.

Amoeba is microscopic (about 0.5 mm across) and you will realize that its surface area is more than adequate. But the mammal has a very low ratio of surface area to volume. The dot represents a cell in the body which is remote from the environment; plainly, a transport system is necessary.
b Increasing an animal's surface area.

3 Use daylight illumination if possible. If not use a lamp, but keep the lamp well away from the microscope to avoid overheating the animal.
4 The specimen may suddenly jerk out of focus, but be patient and focus up and down on the different parts such as limbs and the main body.

Q1 Can you see any streaming of particles?

Q2 Do the streams flow in the same direction?

Q3 Can you see any pumping organ pushing fluid along channels?

Q4 Name two mass-flow systems that you can see.

Gammarus, Asellus, and *Daphnia* are members of the class Crustacea. Their skins are thickened in several places by horny material, but elsewhere the skin is much thinner. They are also active animals.

Q5 Can you relate these facts to their possession of a blood system?

There are other aquatic creatures called planarians which are similar in size to the three crustaceans. Their bodies are thin and leaf-like, their skin is naked, their movements sluggish, their intestines have many branches, and they have no blood system.

Q6 Can you suggest how they survive without one?

14.3 What is blood?

If a few cm³ of mammalian blood are placed in a high speed centrifuge and spun for two or three minutes, the blood separates into two layers. At the bottom of the tube is a dark red layer, and above it a pale yellow fluid. The yellow fluid is the blood *plasma*. It contains water; many ions, including those of sodium, chloride, calcium, and phosphate; and soluble proteins, sugars, and carbon dioxide among many other solutes.

The red layer consists of cells, some types of which are shown in *figures 196* to *198* (pages 215–6). You can examine cells from a sample of fresh blood.
1 Take a clean microscope slide and place a small drop of human blood on it as shown in *figure 195*.
2 Use a second clean microscope slide to draw the drop into a thin film of blood covering much of the slide.

Figure 195
Making a smear of blood.

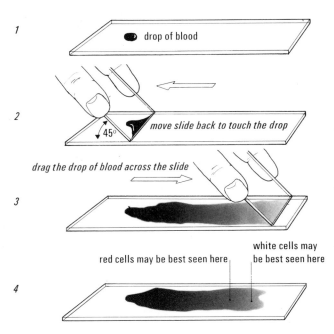

1   drop of blood

2   45°   move slide back to touch the drop

drag the drop of blood across the slide

3

red cells may be best seen here | white cells may be best seen here

4

3   Wave the slide in the air for two or three minutes until the film is dry.
4   Lay the slide on the bench and add one or two drops of Leishman's stain. Add a similar quantity of buffered, distilled water and rock the slide to mix. Leave for five minutes.
5   Wash off the surplus stain under a gently running tap. Shake off the surplus water and wave the slide to dry, or dry it gently over an electric lamp.
6   Examine the smear under the microscope and record the different types of cells you can see.

14.31 The red cells

Most of the cells in the blood are of this type. Their shape is shown in *figure 196*. They are circular, thin, and flexible, with a diameter of about 8 μm (0.008 mm). Each side is hollowed so that the cell is thinner at the centre of the disc than at the edges.

Their red colour is due to an iron-containing pigment called *haemoglobin*, inside the cells, although when seen singly they look rather yellowish in colour.

Since the thin envelope of the red blood cell encloses largely haemoglobin, this substance must have an important role to play in blood circulation. Consider the following facts:
1   Haemoglobin (or a similar compound) is present in the body fluid of most animals whose metabolic rate (energy turnover) is high, except for:

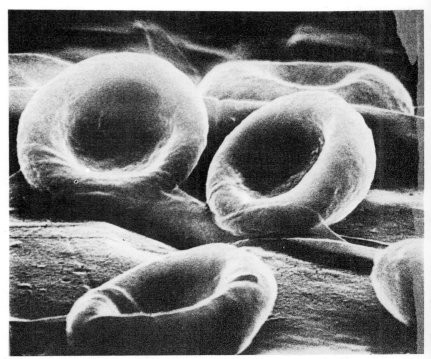

Figure 196
Human red blood cells, × 6000
approximately; taken with a
scanning electron microscope.
There are about 4 500 000 per mm^3
of blood. Red cells live for about
twelve weeks. Dead ones are
removed at the liver and spleen,
new ones being produced in the
bone marrow. The mature cell is
unusual in lacking a nucleus,
though one is present during
development.
Photograph from Nilsson, L. (1974)
Behold man Harrap.

a   insects, whose bodies are penetrated by enormous
     numbers of fine air-conducting tubes or *trachea*;
b   micro-organisms (for example *Amoeba*);
2   One or two species of aquatic animals (for example the
water flea, *Daphnia*) which produce haemoglobin in their
blood only when the oxygen content of their surroundings
falls below a certain value.

This evidence links haemoglobin with the problem of
delivering oxygen to cells. In fact a molecule of
haemoglobin combines with oxygen where it is plentiful
(skin, lungs, and gills), and dissociates from it where the
concentration of oxygen is low. Thus haemoglobin
becomes oxyhaemoglobin at a surface that absorbs oxygen,
but reverts to haemoglobin and free oxygen at a surface
where oxygen is required and used (muscle, brain, gland,
etc.). Each of the four iron-containing *haem* groups in the
haemoglobin molecule can combine with a molecule of
oxygen. Thus at the lung surface (oxygen plentiful):

$$\text{`Hb'} + 4O_2 \rightarrow \text{`Hb'}O_8$$

At the tissue where oxygen is required:

$$\text{`Hb'}O_8 \rightarrow \text{`Hb'} + 4O_2 \qquad (\text{`Hb'} = \text{haemoglobin.})$$

Q1   Can you suggest how the flattened shape of the red cells,
     their small diameter, and their flexibility each contributes
     to their efficiency?

14.32 The white blood cells

Figure 197

Figure 197
Three white blood cells.
× 2000 approximately.
Most white cells are more or less
spherical at rest but change in
outline when active. There are
about 5000 per mm³ of blood but
the number varies, increasing
greatly during an infection of the
body.
*Photograph, Freeman, W. H., and
Bracegirdle, B. (1966) An atlas of
histology, first edition, Heinemann.*

In a stained smear of blood these cells appear blue. They
are far fewer in number than red blood cells and each has a
nucleus (see *figure 197*).

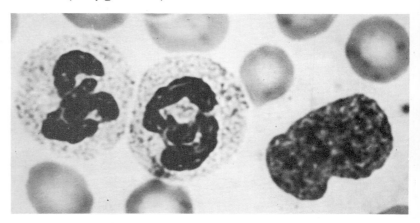

These cells are capable of moving out of the thinnest blood
vessels and a clue to their function has been given in the
caption to *figure 197*. See also colour *plate 9*. A pimple or
spot on the skin is caused by the growth of bacteria. The
white matter (*pus*) in the spot consists of these bacteria and
large numbers of white blood cells. As this pus is
discharged and the wound heals, the bacteria and white
blood cells become fewer in number.

Q2 What do you suggest may be the function of the white
blood cells?

14.33 Platelets

Figure 198
Blood platelets, × 2000
approximately.
*Photograph, Freeman, W. H., and
Bracegirdle, B. (1966) An atlas of
histology, first edition, Heinemann.*

You are very unlikely to see these very small particles in
your own blood smears. They are about 1 μm in diameter

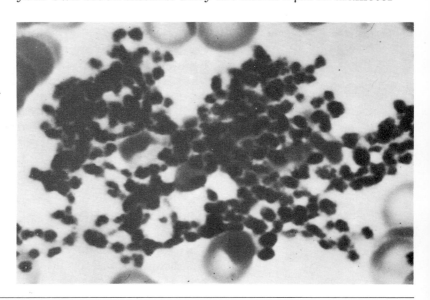

and an irregular shape. You will know that the process of blood *clotting* is literally of vital importance, otherwise you would lose a lot of blood from even the smallest of wounds. When a part of the body is damaged and blood vessels are cut, the platelets help to form a tangled network of *fibrin* threads which form the basis of the clot.

Q3　At this point try to write a summary of the functions carried out by the blood.

14.4 Circulation

In the Background reading, at the end of this chapter, you can read how the understanding of circulation emerged in the seventeenth century, through the work of William Harvey. You can repeat one of his best known experiments yourselves.

Men are better subjects than women, whose thicker fat layer beneath the skin obscures the veins. Having found a suitable subject carry out the following procedure (see also *figure 199*).

Figure 199
Investigating blood flow in a vein.

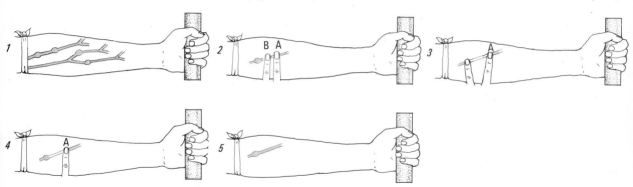

1　The subject should clench his fist or grasp a round stick. The arm should be bared and a moderately tight bandage wrapped around the upper arm above the elbow. This will have the effect of making the blood vessels of the forearm more prominent. (Stage *1* in *figure 199*.)
2　Select a large vein on the inner side of the forearm and place your two first fingers at A and B (stage *2*).
3　Move the finger at B away from A, sliding it towards the upper arm (stage *3*).

Q1　What do you now observe about the vein between the two fingers?

4　Now release the finger you have just moved, leaving finger at A still exerting pressure (stage *4*).

Q2 What do you now observe about the vein above the finger at A?

5 Now release the finger at A (stage 5).

Q3 What do you notice occurs in the vein above A?

Q4 What do you deduce about the direction of flow of blood in the vein?

6 Repeat 2 above and this time move the finger at A away

from B. Maintain the pressure with finger at B and then release A.

Q5 Does this confirm your answer to question 4?

Q6 Do you notice any bulges in these long veins? If so, what do you think they represent?

In this investigation you may have observed that the vessel is clearly divided into sections. If the junction between two such sections were opened up you would find structures like those shown in *figure 200*. These are called 'semi-lunar' or 'pocket' valves.

Figure 200
Valves in a vein.

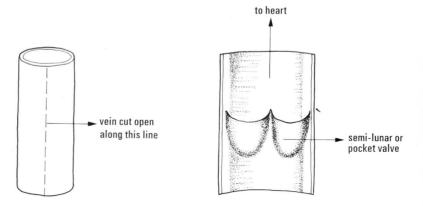

14.5 The plan of a circulatory system

Figure 201 is a plan of the flow of blood in a typical mammal. Colour *plate 10* shows the blood system of man (much simplified). Study these drawings and then answer the following questions.

Q1 Imagine yourself to be a red cell in the main vein of one leg. Which vessels, heart chambers, and other structures would you pass through, in sequence, in order to arrive once more in the same leg?

Figure 201
A plan of the blood circulation in a mammal.

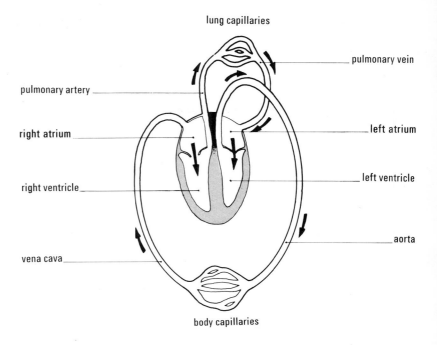

Q2 How many times did the red cell pass through
a the lungs,
b the heart?

Q3 In what ways is blood changed when it passes through the lungs? (See Chapter 5.)

14.6 The heart

The mammalian heart is a beautiful and complex pump which keeps the blood moving around the body. To understand its function it is necessary to first examine its structure.

Examining the external structure of a sheep's heart
1 Note the general shape.
2 Look at and feel the thick, rubbery tubes, the arteries.
3 Distinguish between the right atrium and the left atrium.
4 Examine the thin-walled veins that pass into the atria.
5 Examine the ventricles – note the thicker flesh on one side. This is the animal's *left* side.
6 Look inside the main arteries.
7 Look inside the main veins.

Q1 What could be the function of the blood vessels running over the surface of the heart?

Examining the internal structure of the heart
Open the heart by means of a long ventral cut down the aorta and along the wall of the left ventricle to the apex of the heart. Follow the dotted line as shown in *figure 202*.

Figure 202
Dissecting the heart.
a The dotted line shows how to make a cut down the ventral side.
b The aorta and left ventricle opened. The dotted line shows a second incision.
c A completed dissection of the left side.

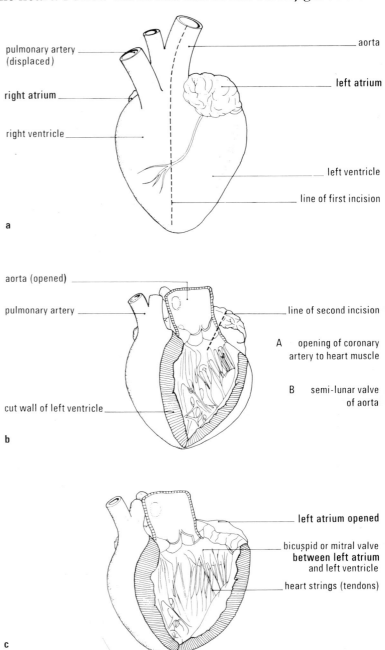

1 Pull the edges of the aorta apart and examine the semi-lunar valves.

Q2 What function do you think they perform?

Living things in action

2 Pull the edges of the ventricle apart and examine the inside of the ventricle.

Q3 What are attached to the valve flaps between atrium and ventricle?

Q4 What do you suggest is the function of these structures?

3 Open the left atrium and its connection with the left ventricle by cutting through the line of second incision shown in *figure 202b.*

4 Pull back the edges of the atrium. Compare the thickness of the walls with those of the left ventricle.

Q5 Why is there this difference in thickness?

5 Note the openings into the atrium of the thin-walled veins. The right side of the heart can be opened in a similar way.

Q6 What do you think might account for the difference in wall thickness between left and right ventricles?

Q7 Why should the aorta have a thicker wall than the pulmonary artery?

Q8 The two ventricles contract and relax at the same time. What comparison can you draw between the blood outputs of each ventricle?

Figure 203
A sectional view of the human heart and its main vessels. *Note:* in *man* the anterior vena cava is usually called the superior vena cava, and the posterior vena cava is called the inferior vena cava. For consistency the *biological* terms have been used throughout this book.

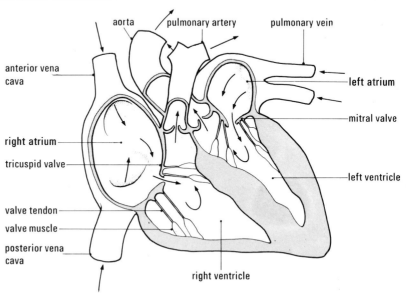

aorta
pulmonary artery
pulmonary vein
anterior vena cava
left atrium
mitral valve
right atrium
tricuspid valve
left ventricle
valve tendon
valve muscle
posterior vena cava
right ventricle

It may be possible for you to see a film loop showing the heart in action. We speak of the heart as a pump, but it would be more correct to think of it as two pumps joined side to side and working in step with one another.

Q9 What sort of blood passes through the heart's left side?

Q10 What sort of blood passes through the heart's right side?

Sounds of the heart
1 Take turns to listen to the sound of the heart by means of a stethoscope.
2 The earpieces should be placed firmly in the ears and the funnel-like end piece placed on your partner's chest to the left of the breastbone. This could be done over a shirt or blouse.
3 Listen carefully for the sounds of the heart while your partner breathes gently.

Heart rate
1 Work in pairs and take the heart rate of your partner by means of the pulse of the radial artery. This artery is in the wrist and the pulse should be detected by the end of the middle finger (see *figure 204*). Your partner should be seated.

Figure 204
Finding the radial artery. Count the pulse on right or left wrist.

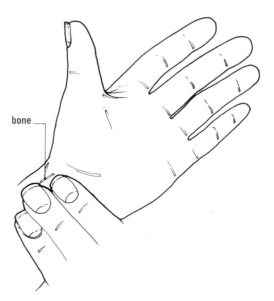

bone

2 Record the pulse rate (beats per minute) of all members of the class on the blackboard.
3 Draw a histogram to show the numbers of individuals at different rates.
4 Find the mean.

Living things in action

Q11 What do you understand by *normal heart rate?*

Making demands on the heart
1 Again work in pairs. One person will be the subject, the other records data.
2 Take the pulse with the subject at rest standing up. Count the number of beats in 30 seconds and multiply by two to obtain beats per minute. Take several readings to check that the pulse is fairly steady.
3 Placing and keeping one foot on a chair, the subject stands alternately on the chair and the floor as quickly as possible for 30 seconds, and then stands still.
4 The subject's pulse is taken again, starting as soon after the exercise as possible. Record how long it takes for the pulse to return to the resting standing pulse.
5 Compare your results with those of other pairs. Try to explain any differences.

Q12 Why does exercise change the pulse rate?

14.7 The blood vessels

Figure 205 shows a capillary bed connected by arterioles and venules with artery and vein. Some of the differences between the vessels have been noted in the figure and we can, therefore, draw up a table of differences on the lines of *table 22*. Examine this and complete it in your notebooks, adding other differences which you may discover.

Differences between

Arteries	Veins
1 Thick muscular walls	Thin layer of muscle in walls
2 Elastic fibres are present in outer coat	Tough fibres are present in outer coat
3	

Table 22

1 Using a microscope, examine a slide showing transverse sections of an artery and a vein. (Compare with *figure 205*.)
2 Look at a dissected mammal and identify some of the main arteries and veins.
3 In a freshly killed and dissected small mammal, the arteries look pale pink and the veins show purple.

Q1 Why should this be their colour?

In preserved dissection specimens, the blood vessels are often emptied and filled with a coloured fluid mixed with a setting agent such as gelatin. The arteries are generally stained red and the veins blue.

Figure 205
A capillary network with
associated arteries, arterioles,
venules, and veins.

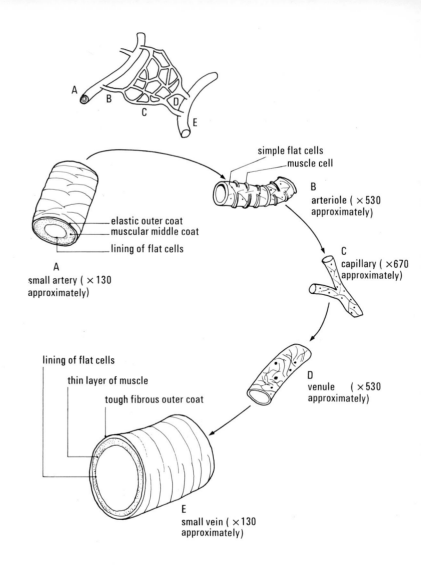

simple flat cells
muscle cell

B
arteriole (×530
approximately)

elastic outer coat
muscular middle coat
lining of flat cells

A
small artery (×130
approximately)

C
capillary (×670
approximately)

D
venule (×530
approximately)

lining of flat cells

thin layer of muscle

tough fibrous outer coat

E
small vein (×130
approximately)

14.71 Looking at capillaries

You can observe the movement of blood in the capillaries, for example in the thin skin between the toes of frogs. Another striking example is in the external gills and tail of frog tadpoles; the tail of a fish provides a further easy view of these tiny vessels and their contents.

These organisms may be set up under the microscope for you to examine. Look at the specimen and focus carefully on the capillaries.

Q2   Can you see any moving objects within the capillaries?

Q3   If so, is the movement regular?

Q4   Do the particles all move in the same direction in different capillaries?

Record your observations by means of notes and simple sketches.

14.8 The lymphatic system

The cells of your body are surrounded by a liquid – *tissue fluid* – which keeps them moist and forms a link with the transport facilities provided by the blood system. Substances required by the cells diffuse from the blood stream into the tissue fluid and so to the cells. Substances produced by the cells (waste substances, hormones, etc.) diffuse back into the blood in the reverse direction.

The spaces between the cells of a man's body hold about 12 dm^3 of such fluid – twice the volume of his blood. *Figure 206* shows how tissue fluid is formed.

Figure 206
Plasma passes out of the arterial end of the capillary and becomes tissue fluid, which can drain back into the capillary at the venous end. Thus, tissue fluid circulates.

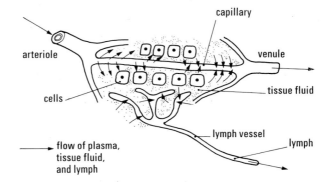

The lymph vessels make up a drainage system for tissue fluid, returning it (as lymph) to the blood. *Figure 207* shows the human lymphatic system. Tissue fluid is being produced at the capillaries all the time and must be returned to the blood system at the same rate.

The lymphatic vessels of the intestine originate in the villi (see Chapter 9) and they are called lacteals because of their milky appearance (Latin: *lacteus* = milky). As long ago as 1622 an Italian, Gasparo Aselli, discovered a series of tubes in the intestine of the dog and he gave them their name. The milky nature is caused by globules of fatty acid and fat absorbed from the digested food in the small intestine.

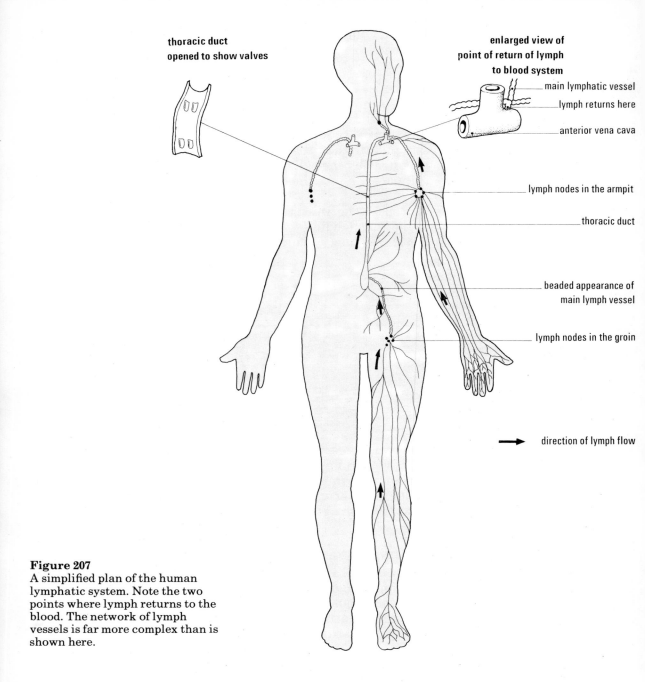

thoracic duct
opened to show valves

enlarged view of
point of return of lymph
to blood system

main lymphatic vessel

lymph returns here

anterior vena cava

lymph nodes in the armpit

thoracic duct

beaded appearance of
main lymph vessel

lymph nodes in the groin

direction of lymph flow

Figure 207
A simplified plan of the human
lymphatic system. Note the two
points where lymph returns to the
blood. The network of lymph
vessels is far more complex than is
shown here.

Background reading

William Harvey and blood circulation

In the Middle Ages the blood was not considered part of a
transport system, with the heart as a pump. Blood was one
of four 'body humours' (the others were *'phlegm'*, *'bile'*, and
'black bile') and the heart was largely the seat of one's
emotional life. It was held that blood moved out of the heart
along the blood vessels in a more or less tidal fashion. As it

Figure 208
An anatomical lecture and
dissection in an Italian university
in the fifteenth century.
*From Singer, S., and Underwood,
E. A. (1962) A short history of
medicine, The Clarendon Press,
Oxford.*

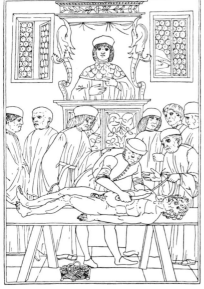

Figure 209
A painting of William Harvey.
*By courtesy of the National
Portrait Gallery, London.*

was consumed by the body, fresh blood was made from food,
drink, and air. Such ideas, the product of scholars in
Ancient Greece and Rome, dominated medicine up to the
seventeenth century. Excessive respect for classical
studies (*scholasticism*) hampered the growth of scientific
method, since the ancients had little time for the
experimental approach. Look at the anatomy lesson in
figure 208.

The students stood around the dissecting table, while above
them the teacher sat on a high chair, often reading aloud
the appropriate passage from the works of Galen (A.D. 130–
200). Below them, a demonstrator used a wand to point
out the structure on a corpse, while a menial carried out
the actual dissection. If observed results varied from
Galen's description, it was explained that the specimen
was abnormal or that the human body had changed since
Galen's time! The possibility that Galen may have been
wrong was never considered.

Around the seventeenth century there was a rise in the
value of experimental science, and among the biologists of
that time was William Harvey, who was born in the middle
of the reign of Elizabeth I in 1578, and died in 1657. He
studied in Padua and returned to St Bartholomew's
Hospital in London, later becoming court physician to
Charles I. In 1628 he published a book called *Anatomica de
motu cordis et sanguinis in animalibus* ('On the anatomy
and motions of the heart and blood in animals'). It has been
called the most important book in the history of medicine.
Harvey was grounded in experimental method at the
University of Padua, and he learned also the value of
studying the same structure in different animals. In his
books he emphasizes the importance of *comparative
anatomy*, both for its own sake and for understanding the
structure of man. He examined the hearts of lizards, frogs,
and fish as well as those of snails, shrimps, and insects.
With insects he used a magnifying glass to detect the
pulsating dorsal heart. Harvey's examination of biological
processes by *experiment* was in the true spirit of Galileo,
who was at Padua at about the same time.

He dissected live animals, and by this vivisection obtained
a much better insight into the working of the blood system
than others who looked only at dead and drained corpses.
He was able to show that the heart is active at 'systole',
when it becomes hardened like a contracted muscle. By
this contraction the blood is expelled from the heart. He
saw the separate movements of the atria and
ventricles and the clear separation of their action to right

and left sides. He proposed that blood flowed from the vena cava into the heart and out through the arteries, and that no *new* blood was required from the liver – as laid down by Galen. Harvey tested his hypothesis with a neat calculation. He estimated that the blood expelled in each beat was 2 oz (56.5 g) and that the heart beat 72 times per minute. Thus in one hour the heart must expel $2 \times 72 \times 60$ oz $= 8640$ oz or 540 lb (249 kg). This is about three times the mass of an average man. Harvey concluded that this mass of blood was far more than could ever be manufactured from any nourishment received, and it was impossible to say where it went to or came from unless one accepted that it went round and round the body. He followed the circulation from the left ventricle and showed how the position of the valves in the veins was now clear, preventing the return flow of blood. The valves in the heart also showed up in their true light in stopping the back pressure of the arteries.

The one link in the chain that he missed was the end of the circulation in the capillaries, for these could only be seen with a microscope. It was Marcello Malpighi, born in 1628, who announced in 1661 the discovery of capillaries in a frog's lungs. It took some 50 years for Harvey's ideas to gain acceptance, though looking at his arguments today we find more meaning in them than anything written on circulation up to that date. Most important, his discoveries opened the way for a new start in the study of living creatures to determine their physiological mechanisms. Only now could one begin to understand respiration, digestion, and other functions. From this start of the circulation of the blood one could now ask basic questions. What did it carry? How did it take up its materials? How and where did it part with them?

15

Maintaining a 'steady state'

15.1 Water balance A. The human body takes in water and loses water at regular intervals throughout every twenty-four hours. It is clear, then, that if our body fluids are to be kept at a constant level, a balance must be struck between input and output.

Consider *figure 210*, showing the daily water balance in the human body. This elaborates on *figure 99* (page 113).

Q1 Which of these quantities will change on the gain and loss sides under very hot conditions?

Q2 Which quantity on the loss side will increase considerably in cold weather? Why should this be so?

The urine contains the largest volume of water lost from the body, and contains a number of substances in solution. We shall see in this chapter that the content of the urine varies according to the regulating control exercised by the kidneys. The amount of water, salts, and other substances, together with waste material, is strictly regulated so that plasma, tissue fluids, and lymph are kept at constant composition. The maintenance of constant internal conditions within the body is called *homeostasis*.

15.11 Production of urine B. The average man produces about 1500 cm³ of urine every day. The exact quantity varies enormously from day to day, and from hour to hour.

Its composition is as shown in *table 23*, overleaf.
below.
(Table underneath this)

You will see that the largest amount of solid matter dissolved in the urine is *urea*. This substance is produced from the excess proteins that are not immediately required for growth, and that cannot be stored. Unwanted protein is changed chemically into a compound which can be stored, plus urea which is eliminated. We can therefore think of urea as a kind of chemical overflow for parts of protein molecules which cannot be stored or put to immediate use. The elimination of such waste products is called *excretion*, and can be defined as follows: 'The process whereby the waste products of the body's metabolism are removed from the body.'

Q3 What processes apart from urine production are excretory?

Figure 210
The daily water balance in an adult.

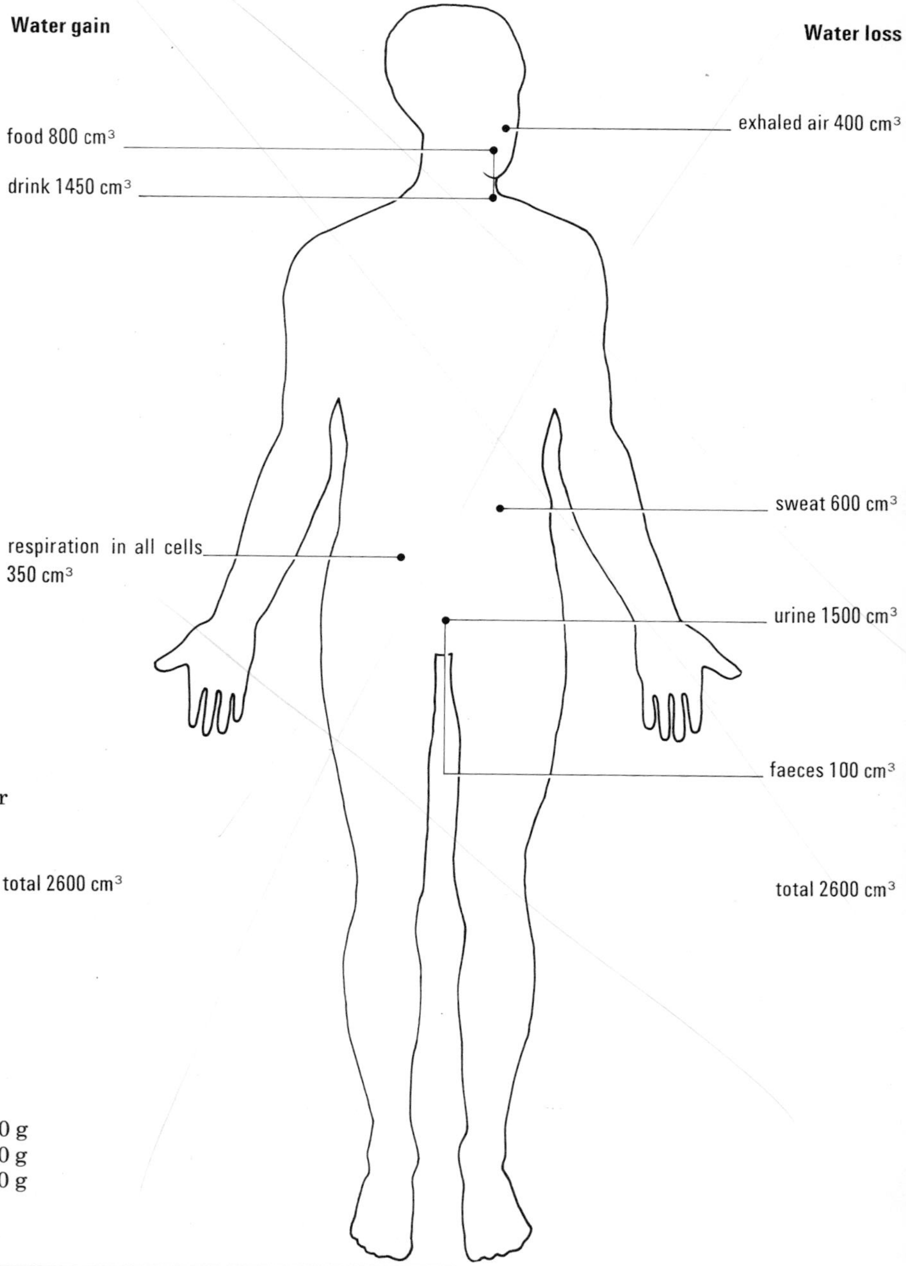

Water gain

food 800 cm³

drink 1450 cm³

respiration in all cells 350 cm³

total 2600 cm³

Water loss

exhaled air 400 cm³

sweat 600 cm³

urine 1500 cm³

faeces 100 cm³

total 2600 cm³

Substance	Amount (grams per 24 hours)
urea	35.0
uric acid	0.8
hippuric acid	0.7
ammonia	0.6
creatinine	0.9
sodium chloride	15.0
phosphoric acid	3.5
sodium	2.5
sulphur	1.2
potassium	2.0

Total quantity of urine	1500 g
water	1440 g
solids	60 g

Table 23

15.2 The urinary system

The urinary system can be seen in *figure 211*. The urine is stored in a muscular sac, the bladder, which leads to the

Figure 211
The human urinary system.

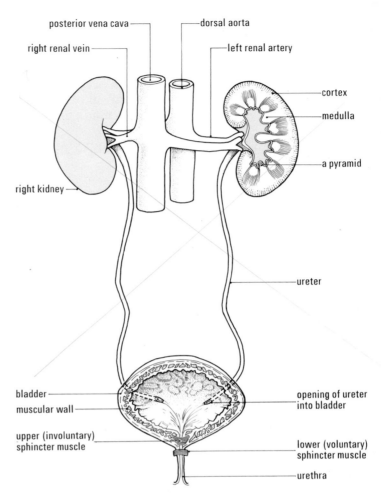

exterior by a short tube, the *urethra*. In the wall of the urethra are two sets of ring-like muscles, or sphincters. These can squeeze the tube to retain the urine, or relax to allow its expulsion. The lower sphincter is voluntary, or consciously controlled, whilst the upper sphincter, at the base of the bladder, is involuntary. It reacts automatically to changes in pressure of the urine in the bladder.

The bladder fills from two tubes, the *ureters*, leading from the kidneys. The kidneys get blood from renal arteries and discharge this blood through the renal veins. You can check the details of *figure 211* by examining a dissection of the urinary system of a small mammal (rat or rabbit).

15.21 The kidney

You will be provided with a kidney.
1 Place it flat on the bench and bisect it longitudinally in the horizontal plane.
2 Remove the top half and examine the various parts of the kidney, its tubes and blood vessels.

3 Identify the enclosing capsule, the cortex, and the medulla (see *figure 211*).

4 Make a sketch of the half kidney showing these parts and also the ureter, the renal artery, and the renal vein.

5 Examine a prepared section of kidney under the low power magnification of a microscope. It helps if the blood vessels have been made more obvious by injection with a coloured fluid.

6 Compare what you see with the photomicrograph in *figure 212*.

Figure 212
A photomicrograph of a section of a kidney in which the blood vessels have been injected with a coloured fluid. The injected material shows up as spots and dark lines. (× 5).
Photograph, W. J. Garnett.

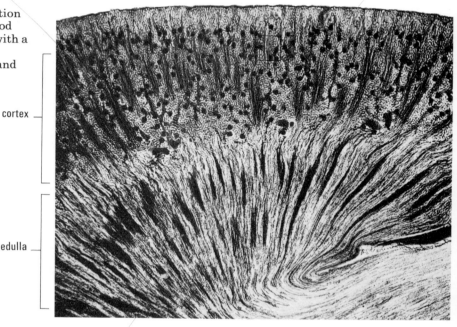

cortex

medulla

7 Now examine a portion of the outer region of the kidney tissue, equivalent to that outlined in *figure 212*, under high power magnification and compare what you see with the photomicrograph in *figure 213*. Inspection of the sections and *figures 212* and *213* shows that most of the material of the kidney is composed of tubular spaces enclosed by cells. Some of these tubes are blood vessels, but the majority are wider and do not carry blood. Careful examination shows that the radiating tubes in the *medulla* lead towards the concave side of the kidney and eventually join each other to form wider tubes, which open into a space from which the ureter extends backwards.

You will not find it easy to interpret this type of complicated tissue from a single slide; it is rather like trying to understand a ball of knitting wool, having cut one single section through the middle of it. A series of sections cut through a whole kidney would be of more help.

Figure 213
A photomicrograph (× 150) of a small area in the kidney cortex, equivalent to the small rectangle outlined in *figure 212*.
Photograph, W. J. Garnett.

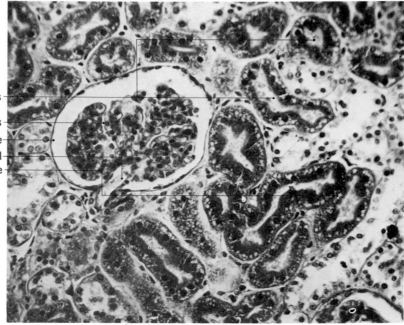

sections of nephron tubes
glomerulus
Bowman's capsule
blood vessel
convoluted tubule

Figure 214 *(below)*
An isolated human nephron (× 8), dissected from the kidney and stretched out so that the convoluted parts have been disentangled. A human kidney consists of about a million nephrons packed together.
Photograph, Professor E. M. Darmady, Tenovus Research Laboratory, Southampton General Hospital.

15.22 The nephron

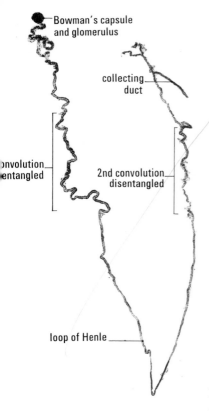

Bowman's capsule and glomerulus

collecting duct

onvolution
entangled

2nd convolution disentangled

loop of Henle

Skilful micro-dissection can be used to isolate these tubes from their neighbours. *Figure 214* shows such an isolated tube which is called a nephron (Greek: *nephros* = a kidney). A human kidney has been estimated to contain a million nephrons. The diagram of a nephron in *figure 215* should help you to interpret *figure 214*.

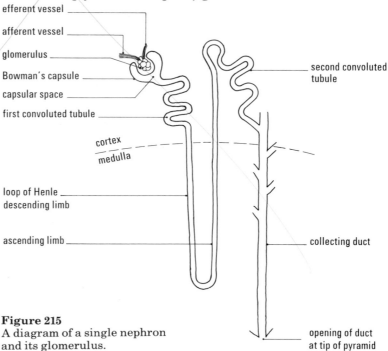

efferent vessel
afferent vessel
glomerulus
Bowman's capsule
capsular space
first convoluted tubule
second convoluted tubule
cortex
medulla
loop of Henle descending limb
ascending limb
collecting duct
opening of duct at tip of pyramid

Figure 215
A diagram of a single nephron and its glomerulus.

Microscopic studies of the kidney reveal the existence of large numbers of blood capillary 'knots' in the outer tissue of the kidney. Each knot can be shown to lie at the end of a nephron, where it is enclosed in a cup of cells called *Bowman's capsule*. In the living kidney, blood enters by the renal artery, which divides repeatedly to supply the separate capillary knots, each of which is called a *glomerulus* (see *figure 216*).

Figure 216
A Bowman's capsule and glomerulus (× 350).

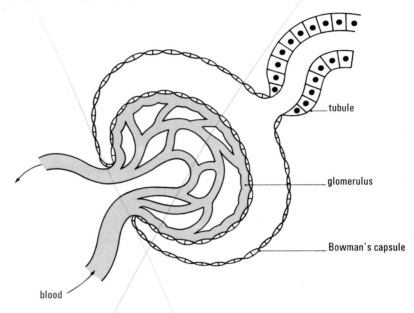

tubule

glomerulus

Bowman's capsule

blood

A further set of capillaries surround the *loop of Henle*. These finally join the renal vein and blood is back in the main circulation.

15.3 How the kidney works

Having looked at the structure of the kidney and its nephrons you can now consider how it functions. Urine is ejected from the open end of the nephron while the other end comes into very close contact with the blood capillaries of the glomerulus (see *figure 216*). The nephron thus appears to be a duct, tapping off certain substances from the blood which are eventually passed to the bladder as urine. You have no evidence yet as to how this is done.

To obtain the evidence of what happens along the nephron we need to analyse its contents at various points. This might seem impossible, since the nephron duct (*tubule*) in its narrowest portions is only 20 μm (0.02 mm) in diameter. Nevertheless, extremely fine pipettes of this diameter have been inserted into those parts of nephrons lying near the surface of the kidney. Samples of the contents of nephrons have then been analysed. By methods which we shall not discuss it is also possible to determine how much liquid is moving along the tubule each minute, that is, the rate of flow. You will presently see that this is not constant along the tubule.

If you look at *figure 216* you will see that the exit capillary from the glomerulus has a narrower bore than the capillary delivering blood.

Q1 Knowing something of the properties of capillaries, can you suggest what effect this will have on the blood pressure in the glomerulus?

The glomerulus and the capsular space are separated by only two layers of cells. These layers act rather like a filter allowing liquid and molecules of a certain size to pass through. Since the filtration occurs under pressure it is called ultra-filtration. Larger molecules are held back just as larger particles are held back by the pore size of a filter. The resulting glomerular filtrate is water plus soluble substances.

The rate of production of filtrate is about 130 cm^3 per minute (over 180 000 cm$_3$ in one day). If you look back to *figure 210* you will see that this is vastly different from the amount of urine passed out of the body in a day. *Figure 217* provides information about the nephron, obtained by sampling techniques mentioned earlier in this section. In *figure 217*, four samples are taken from the nephron and their contents are designated in the drawing. In order to simplify matters, you are not told exactly how much there is in any particular component. Instead, there is an index figure or letter which gives relative information.

Remember these two points:
1 The flow-rate index shows how much fluid is passing a point at a given time. The fluid is largely water and a fall from 100 to 20 must indicate a decrease of 80 per cent, so the substances present in solution must have become 5 times more concentrated.
2 If the concentration index of a solute changes, this must be because of one or more of the following factors:
a whether the substance was ever filtered out of the plasma;
b a change in flow-rate index;
c re-absorption of solute by the nephron.

Now consider the four samples in *figure 217* and answer the following questions.

First sample (compare with the plasma):

Q1 Why is the protein index nil?

Figure 217
A simplified diagram of a nephron and part of its blood supply. The samples show what is happening at various points in the tubule.

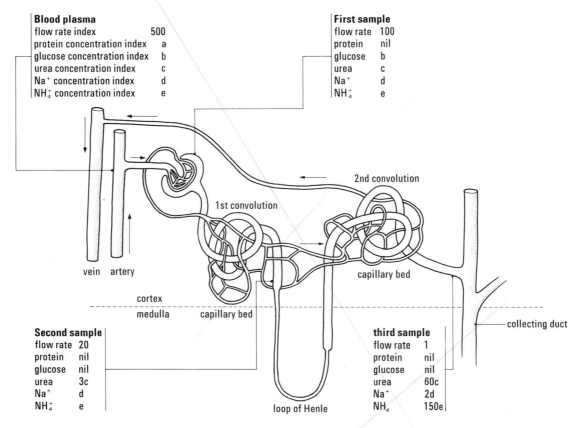

Blood plasma
flow rate index	500
protein concentration index	a
glucose concentration index	b
urea concentration index	c
Na$^+$ concentration index	d
NH$_4^+$ concentration index	e

First sample
flow rate	100
protein	nil
glucose	b
urea	c
Na$^+$	d
NH$_4^+$	e

Second sample
flow rate	20
protein	nil
glucose	nil
urea	3c
Na$^+$	d
NH$_4^+$	e

third sample
flow rate	1
protein	nil
glucose	nil
urea	60c
Na$^+$	2d
NH$_4$	150e

2nd convolution

1st convolution

vein   artery

capillary bed

cortex
medulla   capillary bed

collecting duct

loop of Henle

Q3 What proportion of the fluid part of the plasma has filtered through?

Q4 How does the concentration of sodium ions, ammonium ions, and urea compare with the plasma?

Second sample (compare with the first sample):

Q5 By what percentage is the flow rate reduced?

Q6 What reason can you suggest for the glucose index being reduced to nil?

Q7 How much more concentrated is the urea?

Q8   Compare the concentration of urea shown, with the probable concentration indicated by the flow rate. What has happened to the difference between them? How much is the difference?

Q9   The concentration indices of sodium and ammonium ions have not changed, according to the sample figures, but compare these with the probable concentrations according to the flow-rate index. It would seem that these two should be 5d and 5e. What has happened to 80 per cent of each ion?

Third sample (compare with the second sample):

Q10   By what percentage has the flow rate reduced?

Q11   How much more concentrated is the urea?

Q12   Is this consistent with the concentration indicated by the flow rate?

Q13   How much greater is the concentration of the sodium ions?

Q14   Is this consistent with the concentration indicated by the flow rate?

Q15   How can you explain the difference between these two values?

Q16   What explanation can you give for the increased concentration of the ammonium ions? It should be 20e but in fact it is 150e!

15.4 The kidney as a 'homeostat'

The samples taken from the nephron show that the functions of the kidney may be summarized as follows:

1   Removal of urea and other nitrogenous compounds likely to be toxic – that is, *excretion*.

2   The complete re-absorption of glucose by the nephron and the partial re-absorption of other salts together with a considerable amount of water – that is, *regulation*.
     The amount of water re-absorbed or lost by the kidney depends on the body's requirements. The vital factor is the concentration of the plasma, which must remain constant. The kidney achieves this by regulating water and salt content.

Q1   How could you deliberately decrease the concentration of your blood?

Q2   What effect would this have on the action of the nephrons in your kidneys?

15.41 Salt balance

E. The table under B)

Table 23 in section 15.11 showing urine contents, and *figure 217*, showing the sampling of nephron fluids, both indicates that salt (sodium chloride) is lost from the body. Sweat tastes salty so it seems that the body loses salt through the skin as well as the kidneys. The level of salt in the plasma and tissue fluids will affect the osmotic pressure so that this daily loss must be balanced by intake of salt. Directly the salt level in the plasma changes, the brain must register this fact.

Intense activity means increased sweating and consequently we feel thirsty. This sensation is produced by an increased osmotic pressure of the plasma and tissue fluids. Thus we respond to the dry mouth and craving for fluid by drinking. There is also a need to replace the salt which is lost at an alarming rate in very hot climates, or in such places as steel foundries and coal mines. Taking salt drinks and eating salt with food are methods of compensation which can avoid heat stroke. Man has always recognized this need for himself and his animals.

Q3   What would be the effect on the kidneys if you drank 1 dm^3 of water when you were not thirsty?

Q4   What would be the effect of drinking 1 dm^3 of sodium chloride solution, which had the same concentration as the blood plasma?

15.5 Other adjustments to the internal environment

Physiological studies in recent years have shown that the cells of the body can only function within narrow limits, and that tiny fluctuations of temperature, osmotic pressure, carbon dioxide, oxygen, and certain ions can completely upset the cell's functions. Cells are surrounded by the tissue fluid and this provides a medium in which to live.

Section 14.8 describes the relationship between the tissue fluid and the plasma. Tissue fluid, the material through

Figure 218
The cell in relation to its internal environment.

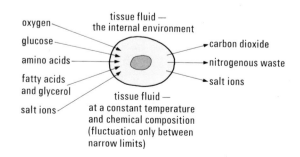

oxygen
glucose
amino acids
fatty acids and glycerol
salt ions

tissue fluid — the internal environment

carbon dioxide
nitrogenous waste
salt ions

tissue fluid — at a constant temperature and chemical composition (fluctuation only between narrow limits) which all materials pass in and out of the cell, can be called the internal environment.

15.51 Carbon dioxide concentration

Respiratory gases are kept constant in the blood stream by the mechanism of breathing. As long ago as 1905, Haldane showed that increasing the concentration of carbon dioxide in the air from 0.03 per cent to 3.0 per cent doubled the respiration rate. Specialized groups of cells are sensitive to

Figure 219
The regulation of carbon dioxide in the blood.

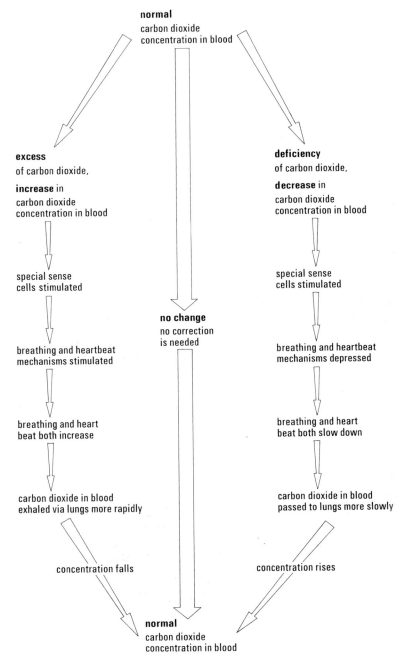

normal
carbon dioxide
concentration in blood

excess
of carbon dioxide,

increase in
carbon dioxide
concentration in blood

special sense
cells stimulated

breathing and heartbeat
mechanisms stimulated

breathing and heart
beat both increase

carbon dioxide in blood
exhaled via lungs more rapidly

concentration falls

no change
no correction
is needed

deficiency
of carbon dioxide,

decrease in
carbon dioxide
concentration in blood

special sense
cells stimulated

breathing and heartbeat
mechanisms depressed

breathing and heart
beat both slow down

carbon dioxide in blood
passed to lungs more slowly

concentration rises

normal
carbon dioxide
concentration in blood

the carbon dioxide levels in the blood, and send messages to the *respiratory centre* of the brain. This centre sends a message along nerves to the breathing organs so that the rate of breathing changes. At the same time, *cardiac centres* of the brain are also alerted and these can modify the heart beat, pumping the blood around the body to eliminate its extra gas more quickly or more slowly.

15.52 Body temperature

In warm-blooded animals, birds and mammals, the body temperature remains fairly constant despite changes in environmental temperatures. Some mechanism in the body must control temperature level. We might ask: 'How hot is the human body?', but this is a vague question with several possible answers. People feel hotter after exercise, but fingers and toes soon become cold on a winter's day.

Body temperature can be measured at different parts of the body, but the commonest is inside the mouth under the tongue. This is unlikely to be affected by external conditions provided the mouth is kept shut. To investigate the body temperature of the members of the class, use a clinical thermometer, put it under the tongue, keep the mouth shut, and read it after one minute. The class results should be collected in a table.

Q1   What first strikes you on examining these temperatures?

Q2   How could you best display them, otherwise than by a table?

Q3   Work out the mean (average) temperature for the class.

Q4   A clinical thermometer has 'normal' marked at 36.9 °C. How does the class mean compare with the national average (so-called 'normal') temperature?

Q5   If you were to repeat the experiment at a future date would you expect to obtain a similar result?

Q6   How does your body temperature compare with that of your surroundings?

Q7   What process is therefore inevitable?

If you examine section 6.2 you will recall that the body continuously liberates energy by respiration within its cells. Heat production must exactly balance heat loss, otherwise body temperature will not remain constant. The major regulatory organ is the skin.

Figure 220
A section through the skin.

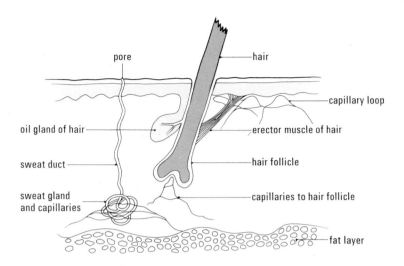

Examine *figure 220* and think about how the structures illustrated help to control body temperature. You can make a simple record of your ideas by copying *table 24* into your notebook and filling it in.

	When body is too hot	When body is too cold
Output from sweat glands		
Attitude of hair (in furry mammals)		
State of superficial blood capillaries ('capillary loops')		
Clothing (of humans) or thickness of fur		

Table 24

How does the body 'know' when to adjust its heating and cooling devices? The detectors are situated in the brain and the skin, but it is the change in temperature of the blood that really determines the switching on and off of the regulator. The brain can detect these small changes in temperature and send nervous impulses to the skin.

The control of the body temperature is absolutely vital, since extreme fluctuations would destroy proteins, inactivate enzymes, and generally disrupt the chemical systems. Slow cooling can cause death – for example, old people living in poorly heated rooms may die from *hypothermia* during winter nights. (See the Background reading for Chapter 3.) Too high a body temperature is equally harmful; a person surrounded by humid air a little warmer than 37 °C could become unconscious in half an hour.

15.6 Maintaining an internal steady state

You have seen that the body's water content, salt content, carbon dioxide concentration, and temperature are all held at fairly constant levels. *Figure 221* illustrates the principle of *feedback*; a change in conditions stimulates a

Figure 221
The principle of control by negative feedback.

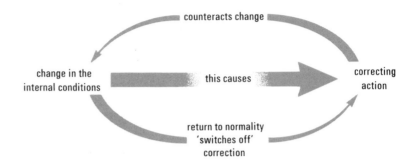

counteracts change

change in the internal conditions — this causes → correcting action

return to normality 'switches off' correction

corrective process which automatically stops when conditions have been completely restored to normal. If you look back, you should be able to identify a few of the body's feedback mechanisms. The maintenance of the internal steady states in an animal is called *homeostasis* (Greek: *homoios* = similar; *statis* = standing). It would be more realistic to use the term 'near-steady', since the inner world of an animal does vary slightly. Some variations, such as the quantity of sugar in the blood, may be fairly extensive. Others, such as the concentration of sodium ions, are regulated with much greater precision. The important point is that such variations are small when compared with the massive physical and chemical changes continually taking place in the body.

Background reading

When things go wrong

Reference to domestic records in the royal household of George III reveals that his urine was sometimes purple; this in turn suggests that his recurring attacks of delirium were the result of a very rare disorder called *porphyria*. Even a normally coloured urine may contain certain unexpected substances. If glucose is constantly present, this indicates the disease *diabetes mellitus*. Diabetes is fairly common and because checking urine for glucose is a frequent test, there is now a very simple sugar test material which (unlike Benedict's solution) requires no boiling.

Again, urine may sometimes contain dissolved protein. If so, this suggests a certain condition called *albuminaria* or one called *toxaemia* (conditions occasionally found in pregnant women).

Kidney failure may happen as a result of accident or disease, and although we can live with only one kidney, when neither is working properly the body is in danger. Until recently the failure of both kidneys meant death within one or two weeks. However, it is now possible, in certain cases, to transplant kidneys, or to provide a machine which can perform the functions of the kidneys. The artificial kidney machine was first thought of in 1913, but it was not properly developed until 1945.

The artificial kidney works by removing unwanted substances across a membrane which allows small molecules (up to about 400 molecular mass) to be separated from larger ones. The differentially permeable membrane is Visking tubing, 0.025 mm thick, and with a surface area of 0.9 m^2. The patient's blood flows on one side of the membrane and on the other side flows a *dialysing* fluid (see below). The artificial kidney models the real kidney in the following ways:

1   Waste products and salts will diffuse out of the blood across the membrane and into the dialysing fluid. This fluid has sodium chloride, sodium acetate, potassium chloride, magnesium chloride, and calcium chloride dissolved in the same concentration as in the plasma of a normal person. The unwanted substances (urea, potassium, and sodium), diffuse out in sufficient quantity to keep the patient well. The process is known as *dialysis*.

2   Water is removed by partially closing the venous tube with a clip which increases the blood pressure to 200 mm of mercury (normal pressure is about 90 mm). Thus, there is ultra-filtration as in the Bowman's capsule of the kidney nephron, which removes about 200 cm^3 of water per hour.

Figure 222
The artificial kidney circuit.

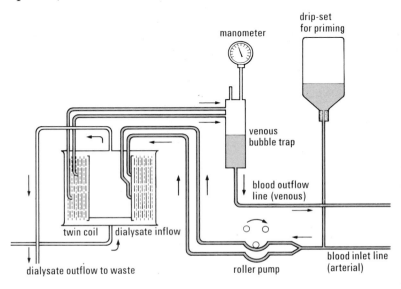

The process is slow, and patients may have to spend all night, twice a week, connected to the machine.

The diagram of the artificial kidney machine (*figure 222*) shows the arrangement of the twin coils of Visking tubing. One coil is connected to the radial artery of the arm. The return flow of blood goes via the other to the cephalic vein of the arm. The bubble trap prevents air bubbles getting into the blood system. The blood is kept moving by a roller pump. The patient settles in bed for the night and is connected to the machine. In the morning, after he has been weighed to ensure that the right amount of excess water has been removed from his body, the patient is disconnected and can go home.

Appendix

Table showing
the composition of some foods

Food	Inedible waste	Energy value		Protein	Fat	Carbo-hydrate	Calcium	Iron	Vitamin A	C	D	Thiamine	Riboflavin	Nicotinic acid
	%	kcal kJ per 100 g of food		%	%	%	mg	mg	µg	mg	µg	mg	mg	mg
									per 100 g of food					
milk	0	65	272	3.3	3.8	4.8	120	0.1	44	1	0.05	0.04	0.15	0.1
yoghurt	0	57	239	3.6	2.6	5.2	140	0.1	39	0	0.02	0.05	0.19	0.1
cheese, cheddar	0	412	1726	25.4	34.5	0	810	0.6	420	0	0.35	0.04	0.5	0.1
bacon, average	13	476	1994	11.0	48.0	0	10	1.0	0	0	0	0.4	0.15	1.5
beef, average	17	313	1311	14.8	28.2	0	10	4.0	0	0	0	0.07	0.20	5.0
chicken, roast	0	184	771	29.6	7.3	0	15	2.6	0	0	0	0.04	0.14	4.9
ham, cooked	0	422	1768	16.3	39.6	0	13	2.5	0	0	0	0.50	0.20	3.5
lamb, roast	—	284	1190	25.0	20.4	0	4	4.3	0	0	0	0.10	0.25	4.5
liver, fried	0	276	1156	29.5	15.9	4.0	9	20.7	6000	20	0.75	0.30	3.50	15.0
pork, average	15	408	1710	12.0	40.0	0	10	1.0	0	0	0	1.0	0.20	5.0
sausage, pork	0	369	1546	10.4	30.9	13.3	15	2.5	0	0	0	0.17	0.07	1.6
cod, fried in batter	0	199	834	19.6	10.3	7.5	80	0.5	0	0	0	0.04	0.10	3.0
salmon, canned	2	133	557	19.7	6.0	0	66	1.3	90	0	12.5	0.03	0.10	7.0
eggs, fresh	12	158	662	11.9	12.3	0	56	2.5	300	0	1.5	0.10	0.35	0.1
butter	0	745	3122	0.5	82.5	0	15	0.2	995	0	1.25	0	0	0
lard	0	894	3746	0	99.3	0	0	0	0	0	0	0	0	0
margarine	0	769	3222	0.2	85.3	0	4	0.3	900	0	8.0	0	0	0
chocolate, milk	0	578	2422	8.7	37.6	54.5	246	1.7	6.6	0	0	0.03	0.35	1.0
jam	0	262	1098	0.5	0	69.2	18	1.2	2	10	0	0	0	0
ice cream, vanilla	0	192	805	4.1	11.3	19.8	137	0.3	1	1	0	0.05	0.20	0.1
sugar, white	0	394	1651	0	0	100	1	0	0	0	0	0	0	0
beans, canned	0	92	385	6.0	0.4	17.3	62	2.1	50	3	0	0.06	0.04	0.5
beans, runner	14	15	63	1.1	0	2.9	33	0.7	50	20	0	0.05	0.10	0.9

Table continued on page 246

Food	Inedible waste	Energy value		Pro-tein	Fat	Carbo-hydrate	Cal-cium	Iron	Vitamin A	C	D	Thia-mine	Ribo-flavin	Nico-tinic acid
	%	kcal	kJ	%	%	%	mg	mg	μg	mg	μg	mg	mg	mg
		per 100 g of food									per 100 g of food			
cabbage, boiled	0	8	34	0.8	0	1.3	58	0.5	50	20	0	0.03	0.03	0.2
carrots	4	23	96	0.7	0	5.4	48	0.6	2000	6	0	0.06	0.05	0.6
lettuce	20	11	46	1.1	0	1.8	26	0.7	167	15	0	0.07	0.08	0.3
peas, boiled	0	49	205	5.0	0	7.7	13	1.2	50	15	0	0.25	0.11	1.5
potatoes, boiled	0	79	331	1.4	0	19.7	4	0.5	0	4–15	0	0.08	0.03	0.8
potato chips, fried	0	236	989	3.8	9.0	37.3	14	1.4	0	6–20	0	0.10	0.04	1.2
tomatoes, fresh	0	14	59	0.9	0	2.8	13	0.4	117	20	0	0.06	0.04	0.6
watercress	23	14	59	2.9	0	0.7	222	1.6	500	60	0	0.10	0.16	0.6
apple	21	46	193	0.3	0	12.0	4	0.3	5	5	0	0.04	0.02	0.1
banana	40	76	318	1.1	0	19.2	7	0.4	33	10	0	0.04	0.07	0.6
orange	25	35	147	0.8	0	8.5	41	0.3	8	50	0	0.03	0.2	0.3
pineapple, canned	0	76	318	0.3	0	20.0	13	1.7	7	8	0	0.05	0.02	0.2
rhubarb	33	6	25	0.6	0	1.0	103	0.4	10	10	0	0.01	0.07	0.3
peanuts, roasted	0	586	2455	28.1	49.0	8.6	61	2.0	0	0	0	0.23	0.10	16.0
biscuits, rich, sweet	0	496	2078	5.6	22.3	72.7	92	1.3	0	0	0	0.12	0.04	1.0
bread, brown	0	237	993	9.2	1.8	49.0	92	2.5	0	0	0	0.28	0.07	3.2
bread, white	0	253	1060	8.3	1.7	54.6	100	1.8	0	0	0	0.18	0.02	1.4
cornflakes	0	365	1529	7.5	0.5	88.0	5	1.1	0	0	0	0.60	1.07	7.0
spaghetti	0	364	1525	9.9	1.0	84.0	23	1.2	0	0	0	0.09	0.06	1.7
coffee, instant	0	156	654	4.0	0.7	35.5	140	4.0	0	0	0	0	0.10	45.0
coffee, ground	0	0	0	0	0	0	0	0	0	0	0	0	0.2	10.0
tea, dry	0	0	0	0	0	0	0	0	0	0	0	0	0.9	6.0
beer, mild draught	0	25	105	0.2	0	1.6	10	0	0	0	0	0	0.05	0.7
fruit cake, rich	0	368	1542	4.6	15.9	55.0	71	1.8	56	0	0.80	0.70	0.3	1.2
rice pudding	0	142	595	3.6	7.6	15.7	116	0.1	96	1	0.08	0.05	0.14	0.2
soup, tomato, canned	0	67	281	0.9	3.1	9.4	18	0.3	46	6	0	0.03	0.02	0.5

Index

cheetah: bones of leg of, 27; running speed of, 22
chemical energy, conversion of food energy into, 89
chemical processes of body, effect of temperature on, 43, 54
Chlorella: uptake of carbon dioxide by, in light and darkness, 153–4
chloride: in alga and pond water, 199; in blood, 213; in body fluids, 113; not essential in plant culture solutions, 197
chlorophyll, 166, 172
chloroplasts, 172–3
cod, osmotic relations in, 189
Commelina, stomata on leaf of, 182
consumers, primary and secondary, 161
Cook, Captain, 109
coronary artery, 220
cress, roots of, 190
Crustacea, blood system in, 213
culture solutions for plants, 196, 197; experiments with, 198–9
custard powder, apparatus for burning, 84

D

Daphnia (water-flea), 212–13; produces haemoglobin only in conditions of oxygen shortage, 215
DCPIP (dichlorophenol indophenol), for estimation of amount of ascorbic acid in food, 109–10
death rates: from all causes, smokers and non-smokers, 79; from lung cancer, all men and male doctors, 80; non-smokers, continuing smokers, and ex-smokers, 81
deer: four types of, 48
defecation, 144
dentition: and diet, 125; permanent and deciduous, 120
desert, aestivation of snails in, 192
diabetes mellitus, 242
dialysis, 243
diaphragm, 69; movements of, in breathing, 67, 71
diet: of American city-dweller and Indian workman, 118; complete ('balanced'), 103; dentition and, 125; roughage in, 144
diffusion: in agar jelly, 210; difference between mass flow and, 211–12; of gases in leaf, 173
digestion, 139; study of, 141–2, 144–7

digestive system, human, 138, 139–41
disease: combined effects of malnutrition and, 116; transmitted by houseflies, 130
doctors: changes in smoking habits of, and in lung-cancer death rates of, 80
drag in movement, 17; in flight, 30, 31; shape and, 33
duodenum, 138, 140, 141

E

elbow joint, 38, 39
electrical energy, conversion of food energy into, 88
elephant: bones of legs of, 7–8, 27; mass and height of, 2, 3; mass and temperature of, 42, 43; skeleton of, 9
elk or moose (*Alces*), 48, 49
Elodea (water-plant), 155, 156
emulsion of fats, 106, 109, 142
energy: conversion of, in living organisms, 91–2; from food, 84–7, 102, 245–6; human requirements for, 113–15; measuring human output of, 90–1; for photosynthesis, 166, 173; release of, by living organisms, 87–90
enzymes (biological catalysts), 138–9; digestive, 139, 141, 142
epidermis of leaf, 169, 170, 171; translucent, 173
ergometer, bicycle, 90–1
ethanol, produced from sugar by yeast, 95
eutrophication, in waters receiving excess of fertilizers, 209
evaporation, transpiration and, 181–2
Everest, Mount, ascent of, 63, 64
excretion, 230; by kidney, 237
'explosive events', in athletics, 97–8
eyes; vitamin A and, 111, 115

F

faeces, 144, 230
fats: absorption of, from gut, 142, 143, 225; in food, 103, 106; percentage of, in different foods, 245–6; test for, 106, 109
feathers, 27, 29
feedback, 242
fermentation, by yeast, 95
fertilizers, chemical, 208–9

fishes: luminescent, 87; osmotic problems of, 188–9; swimming by, 34–6; vitamin D in livers of, 111
flight, of birds, 27–32
flow rate index, for kidney, 235, 236
fluoride, and dental caries, 131–3
food, 102–7; composition of different kinds of, 245–6; energy from, 84–7, 113–15; minerals in, 112–13; oxidation of, 87, 91, 93; plants as source of all, 161; prospects of greater efficiency in production of, 178; tests for different kinds of, 108–9; vitamins in, 109–12; water in, 113, 230; water produced in metabolism of, 230; world problem of, 115–18
food chains and webs, 161, 173–5
foxes: three types of, 49, 50
frog: evaporation of water through skin of, 193; observing blood movement in capillaries of, 224; sequence of movements of, in jumping, 26
fuel: energy from combustion of, 91–2; food as, 84, 102

G

Galen, 227
gall bladder, 142
Gammarus (freshwater shrimp), 212–13
gas analysis, 58–9, 156
gas exchange by living organisms, 60, 148, 149–50, 152–3
gastric juice, 139, 145–7
Geiger counter, 153, 154
gerbil, mass and surface area in (adult and baby), 44, 45
gills, excretion of salt through, 189
giraffe, running speed of, 23
gizzard, of locust, 127
glasshouse crops, carbon dioxide for, 175–8
glomerulus of kidney, 233, 234, 235
glucose, 103; breakdown of, by living cell, 92; breakdown of starch to, 135–7, 141; in cellulose tubing, 134–5; effect of, on muscle fibres, 92–3; formation of starch from, 164–5; labelled, rats produce labelled carbon dioxide from, 85; reabsorbed in nephron, 237; storage of excess, as glycogen, 143; structure of molecule of, 104, 105, 164
glucose–1–phosphate, 164
glycogen, 143

goosegrass (*Galium*), from different habitats, 16
grass: percentage of energy in crop of, converted into meat by bullock, 174–5
grasshopper, production of sound by, 88
greyhound, running speed of, 22
growth: on inadequate food, 102; proteins for, 107
guard cells of stomata, 169, 170
gut (alimentary canal), 134, 138; muscles of wall of, 139–41

H

habitat of plants, and form of growth, 16
haemoglobin, 214–15
Hales, Stephen, 203
Harvey, William, 217, 227–8
heart: hypothermia for surgery on, 56; sounds of, 222; structure of, 219–21; in training for long-distance races, 99
heart beat, rate of, 222, 228; carbon dioxide content of blood and, 239–40; effect of exercise on, 223; in hibernation, 55
heat energy, conversion of food energy into, 43, 88–9
hedgehog, hibernation of, 55
Helmont, Jean-Baptiste van, 157–8
herbivores, 125, 161; breakdown of cellulose in gut of, 144
hibernation, 192; hypothermia in, 55
high altitudes, acclimatization to, 63, 100–1
hip joints, X-rays of, 40
hogweed (*Heracleum*), stem of, 12
homeostasis, 229, 242
homeostat, kidney as, 237–8
homoiothermic (warm-blooded) animals, 42, 53, 240
Hopkins, F. G., 110–11
horse: different breeds of, 25; foot of, 23; legs of, 7–8, 27; problems of support in, 2; sequence of movements of, in jumping, 26; skeleton of, 9
housefly, feeding by, 128–30
humans: digestive system of 138, 139–47; foot of, 14; heart of, 221–3; hypothermia in, 54–5, 241; joints of, 38–9; muscles of, used in running, 21; output of energy by, 90–1; respiratory system of, 57–9, 67–78; skeleton of, 9, 19; teeth of, 119–22

hydrogen: added to carbon dioxide in photosynthesis, 165; in carbohydrates, 103, 105; in fats, 106; in proteins, 107
hypothermia, 53–6, 241

I

Ingen-Housz, John, 159–60
insects: feeding by, 126–30; 'heart' of, 227; luminescent, 87; tracheae of, 215
intercostal muscles, 69, 70
intestine, 138, 139; absorption of water in, 144; villi of, 142–3
iodine, in thyroid gland, 113
iron: content of, in different foods, 245–6; in haemoglobin, 215; human requirements for, 114; in red blood cells, 113
isotopes: of carbon, 85, 153, 204–6; of oxygen, 165; of sulphur, 200

J

jaws, muscles of, 124–5
joints, in movement, 18, 37–40
joules and kilojoules, 86
jumping movement, 18, 26–7

K

kidney, 231–4; artificial, 243–4; failure of, 243; of freshwater and marine fish, 189; as a homeostat, 237–8; how it works, 234–7
Knop, Wilhelm, 197
kob (*Odonata*), 2
kwashiorkor, disease caused by lack of protein, 102, 115

L

lacteals, 143, 225
lactic acid, produced in muscles by anaerobic respiration, 94, 96, 98
Lawes, J. B., 207
lead, in soil and plants, 196
leaves: site of photosynthesis in, 172–3; structure of, (internal) 170–2, (surface) 168–70; structure of, and transpiration, 182; testing for starch in, 162–3; transport of food from, 203–6; variegated, 163, 167
legs, 3; bones of, 'pillars' or 'zigzags', 7–8; length and thickness of, in relation to size of animal, 3, 4
lettuces, grown with and without carbon dioxide enrichment, 177
Liebig, Justus von, 207
ligaments, 23, 37, 39

light: and absorption of carbon dioxide by plants, 150–3; and presence of starch in leaves, 163
light energy: amount received per square metre per year in temperate latitudes, 173–4; conversion of food energy into, 87; percentage of, trapped by plants, 174; in photosynthesis, 166, 173
Lind, James, on cure of scurvy, 109
liver, 143, 215
locust, feeding by, 126–8
luminescent organisms, 87
lungs: air pressure in, 73; capacity of, 74–5; casts and bronchograms of, 67, 68; cigarette-smoking and cancer of, 79–81, 83; fine structure of, 76–7; gas exchange in, 60, 67–72, 77–8
lymphatic system, 143, 225–6

M

Magendie, François, 109
magnesium: in alga and pond water, 199; in bones and cells, 113; in soil and plants, 196
malnutrition, 115–18
Malpighi, Marcello, 204, 228
mammals: body temperature and mass of, 42; plan of blood flow in, 218–19; size and support in, 2–9
mandibles of locust, 126, 127
manganese, in soil and plants, 196
manuring of crops, 206–9
marasmus, in underfed baby, 102
Martin, Alexis St, 144–7
mass of animal: and shoulder height, 4; and surface area, 43–6; and temperature, 42–3; and volume, 6
mass flow, for transport in animals, 211–12
mastication, 119
membranes (permeable, impermeable, and differentially permeable), 183–4
mesophyll of leaf, 171, 173, 182
microvilli, 143
milk, vitamins in, 111
minerals: in alga and pond water, 199; in food, 103, 112–13; in plants and soil, 196
models: ball and rod, of glucose molecule, 105; bell-jar and syringe, of thorax, 72; 'container', of animals, 5; cube, to show relations of linear dimensions and volume, 4–5, and of surface area and volume,

roots of plants, 9–10; absorption of water by, 190–1; in moist and dry soils, 193; transport of liquid from, 201–3

Rothamsted experimental agricultural station, 207; Broadbalk wheat experiment at, 207–8

roughage, in diet, 144

running movement, 18, 21–5

S

Sachs, Julius von, 197

saliva, 119, 121; breakdown of starch by, 135–7; of housefly, 130; of locust, 127

salmon, migration of, 193–4

sand dunes, effect of mineral supplement on plants growing in, 195–6

scurvy, 109, 112

seal (*Halichoerus*), 3

shape of animal: and drag in flight, 33; and movement in water, 34–5; and resistance to wind, 14; and stability, apparatus for testing, 15; and way of life, 3

sheep, teeth of, 123, 124

shivering, 54

shoulder joint, X-ray of, 38

shrew, 42, 43, 45, 53

skeleton: of animals of different sizes, 9; human, 9, 19; and thrust in movement, 36

skin, in regulation of temperature, 240–1

smoking, 78; of cigarettes, association of, with lung cancer, 79–81, 83, and with other diseases, 79, 83

snails in desert, aestivation of, 192

sodium: in alga and pond water, 199; in blood, 213; in body fluids, 113; in soil and plants, 196

sodium chloride (salt), balance of, 238

soil, mineral contents of plants and, 196

sound energy, conversion of food energy into, 88

space travel, oxygen for, 65–6

spectrum, absorption: of chlorophyll, 167

sphincters, 140; anal, 144; of bladder and urethra, 231; pyloric, 138, 140–1

spleen, 213

starch, 104, 105, 164; breakdown of, to glucose, 135–7, 141; in cellulose tubing, 134–5; in food,

103; formed from glucose, 164–5; test for, 104, 108; testing leaves for, 162–3

stems of plants, 9–10; internal structure of, 12, 202–3; resistance to sideways forces by, 11–12

stomach, 139, 140–1; fistula of, 144–7

stomata of leaves, 169–70, 173; at different times over 24 hours, 182; loss of water vapour through, 182

stride, definition of, 19

sucrose: onion cells in solution of, 186–7; tracing spread of radioactive, in shoot of plant, 205–6

sugar beet and sugar cane crops, trapping of light energy by, 174

sugars: in food, 103; in phloem, 206; test for reducing, 104, 108; and tooth decay, 122; *see also* glucose

sulphate: oxygen and uptake of, by plants, 200

sulphur: in proteins, 107; radioactive, 200

sulphur dioxide, poisonous to tomatoes, 177

surface area: of alveoli in lungs, 77; and mass, 43–6; of mesophyll in leaf, 182; of villi in gut, 142; and temperature, 42–3; and volume, 210–12

surgery, hypothermia in, 55–6

sweat: salt in, 238; water in, 230

swimming, by fish, 34–6

synovial membrane, synovial fluid, 37, 39

synthesis, 166

T

tax, on tobacco, 83

teeth, 119–21; calcium required for, 102; prevention of decay of, 121–2, 131–3

temperature of body: keeping it constant, 41–2, 240–1; 'normal', for humans, 88; problems on, 52–3; regulation of, 53–6; size and, 42–51

tench, coat of mucus on, 189

tendons, in movement, 18

thrust in movement, 17; skeletons and, 36

thiamine (vitamin B group), 110, 112; content of, in different foods, 245–6; human requirements for, 114

thoracic duct, 226

thorax, human, 69; bell-jar and syringe models of, 72

tissue fluid, 225; relation between blood plasma and, 238–9

Titanic, deaths from hypothermia at sinking of, 55

tobacco, advertising of, 82–3

toxaemia, 242

tracheae, of insects, 215

training, in athletics, 97–9

transpiration in plants, 180, 201; and evaporation, 181–2

trees, support of, 10

trolley, experiments on movement with, 17, 18

turbulence of air, and flight, 31

U

ultrafiltration, in kidney, 235; in artificial kidney machine, 243

ultraviolet light, and vitamin D, 111

urea, 229, 230

urinary system, 230–1

urine, 229–30; output of, per unit weight of body, for human and freshwater fish, 189

V

valves: bicuspid or mitral, in heart, 220, 228; semi-lunar or pocket, in aorta, 220, in thoracic duct, 226, and in veins, 218, 228

veins, human: differences between arteries and, 223–4; studying blood flow in, 217–18

veins, of leaf, 171, 172, 173

vena cava, 219, 220, 221, 231; return of lymph to blood system in, 226

ventricles of heart, 219, 220, 221; observed by Harvey, 227, 228

villi of intestinal wall, 142–3, 225

vitamin A (retinol), 111, 112; blindness from lack of, 115; content of, in different foods, 245–6; human requirements for, 114

vitamin B, *see* nicotinic acid, riboflavin, thiamine

vitamin C (ascorbic acid), 109, 112; contents of, in different foods, 245–6; human requirements for, 114

vitamin D (calciferol), 111, 112; content of, in different foods, 245–6; human requirements for, 114

vitamins, 103, 109–12, 114, 115

vocal cords, 88

W

walking movement, 18

water: absorption of, in large intestine, 144, and in nephron, 237; daily balance of, in man, 103, 113, 229, 230; effect of fluoride in, on dental caries, 131–3; movement of, through membranes, 183–4 (*see also* osmosis); percentage of, in different organisms, 119; of river, oxygen in, over 24 hours, 154–5; splitting of, in photo-synthesis, 165, 173; support from, for aquatic animals, 8, and for plants, 15; uptake and output of, by plants, 179–82; van Helmont's view that plants must consist mainly of, 157–8

weasel (*Mustela*), 3

whale: could it have legs? 8; skeleton of, 9

wheat: breeding for stronger stems in, to support heavier heads, 16; effect of mineral manures on growth of, 208

white blood cells, 216

wind: effect of, on different plants, 13

wind tunnel, for studying flight, 31

wing of bird, 27, 29; forces acting on, 30–1

woundwort, hedge (*Stachys*), stem of, 12

X

X-ray photographs: of chest, 71; of gut, 140; of joints, 37, 38, 40; of lung, 67, 68

xylem of plant stem, 202, 203

Y

yak (*Bos grunniens*), 49

yeast, anaerobic respiration in, 94–5

Z

zinc, in soils and plants, 196